WHO DARES WINS

WHO DARES WINS

COMBAT CAPABILITY GENERATION FOR SPEED, ADAPTATION, AND BATTLEFIELD DOMINANCE

Brian C. Stehle

This publication contains the opinions and ideas of its author and is designed to provide useful advice to understand the concepts of Combat Capability Generation (C2G).

Printed in the United States of America

Cover Design by Brian C. Stehle

ISBN:	Hardcover	978-1-7357272-3-3
	Softcover	978-1-7357272-2-6
	eBook	978-1-7357272-4-0

DEDICATION

This work is dedicated to those who have ever worn or wear the uniform, to those who never will but are protected by those who do, and to those who will inherit the world that our decisions shape.

May this framework make conflict less necessary — and peace more resilient.

DISCLAIMER

The opinions expressed in this book are those of the author alone and do not represent the official position, policies, or views of the Military Services or any other agency of the United States Government.

This work is intended for conceptual, educational, and strategic discussion only. Nothing in this publication constitutes operational guidance or classified direction.

The author began developing the ideas contained in this book on November 20, 2025 shortly after the release of the Department of Defense Acquisition Transformation memorandum. Any correlation in timing is reflective of a broader institutional inflection point, not a coordinated effort.

This work was supported by modern research and productivity tools, including generative language systems, used strictly as drafting and organizational aids. Their use does not constitute endorsement of any specific product or brand.

AUTHOR'S NOTE

Over my career as a Soldier, aviator, commander, and program manager, I saw a simple truth: our people adapt faster than our processes. Soldiers solve problems in real time, under pressure, with creativity and teamwork. Too often, the systems meant to support them move at a different speed.

This book is an acknowledgment that the world has changed—and our approach to modernization must change with it. Technologies evolve faster, threats adapt faster, and opportunities appear faster than our legacy models can absorb. We owe our Soldiers a system that adapts as quickly and responsibly as they are asked to fight.

C2G is my attempt to offer a clearer, more agile way forward using principles we already trust: Effects-based thinking, mission command, competition, modularity, and continuous learning. It focuses on alignment, speed, and coherence—not complexity. And above all, it keeps the Soldier at the center.

This book is written for senior military leaders, policymakers, Congressional leaders and staff, defense industry executives, and strategists who are responsible for making consequential decisions about how the nation generates combat capability. It is intended for those who operate at the level of systems, authority, and governance—where time, risk, and legitimacy matter as much as technology or budget. While the concepts apply broadly, this is not a tactical manual or a technical acquisition guide. It is a framework for leaders who must balance speed with discipline, innovation with accountability, and adaptation with ethical responsibility in an era where delay itself has become a strategic risk.

If this framework helps a leader think differently, make a decision, or deliver capability faster, then the effort was worth it. Our nation is capable of providing this support. Our Soldiers deserve a system that matches their pace.

— **Brian C. Stehle**

ACKNOWLEDGEMENTS

This book came together over a short period of time, but it did not emerge quickly. It reflects decades of shared experience across military service, acquisition, industry, and joint problem-solving environments. While the framework is presented as a unified concept, it is built from countless conversations, lessons learned, and moments of clarity provided by others.

I am grateful to the mentors who invested time and trust in developing my judgment—leaders who modeled disciplined decision-making, ethical responsibility, and the courage to act when certainty was unavailable. Their influence shaped how I think about capability and speed in addition to leadership itself.

I acknowledge Mike Harris and Tim Hope for their thoughtful feedback during the development of this work, which helped sharpen several core arguments. I also thank the colleagues, peers, and subordinates who challenged ideas, exposed blind spots, and pushed back when assumptions deserved scrutiny. Many of the strongest elements of this framework emerged not from agreement, but from thoughtful disagreement among professionals equally committed to the mission. Those exchanges strengthened the ideas presented here and grounded them in reality.

Most importantly, I acknowledge the Soldiers and civilians whose adaptability continues to outpace the systems around them. Their creativity, resilience, and sense of responsibility remain the foundation of any credible capability. This work is offered in service to them, with the hope that it contributes—however modestly—to a system that better matches their pace and professionalism.

Any errors, omissions, or oversights remain my responsibility alone — and like the framework it proposes, they are not endpoints, but starting points for learning, iteration, and improvement.

TABLE OF CONTENTS

EXECUTIVE SUMMARY

The modern security environment operates at a speed that has shattered conventional military timelines, moving faster than the measured, deliberate processes many defense institutions rely upon. The core challenge is the **"Dinosaur Imperative,"** recognizing that overwhelming strength without agility is temporary; success in one era is not insurance for the next. Static planning and rigid requirements generate obsolescence before delivery, creating strategic vulnerability through delay.

This book introduces **Combat Capability Generation (C2G)**, a unifying framework designed to evolve these systems by aligning needs, Effects, innovation, governance, ethics, and execution to achieve the necessary speed. C2G shifts the enterprise away from platform-based thinking toward **Effect-Based Thinking**. Instead of asking "What system should we build?", C2G asks: **"What change (Effect) is required given a situation, and how can it be achieved faster?"** This approach aligns the enterprise with the Observe and Orient elements of the OODA Loop, ensuring relevance and focus.

C2G creates an intelligent, adaptive structure built on disciplined speed and logic. Key components include:

• **The Capabilities and Effects Registry (CER):** This is the central, living "brain" of C2G, acting as the single repository of required Effects, ensuring unity of mission, managing capability gaps, and preventing unnecessary duplication of effort. Effects are owned by the mission, not by traditional domains or branches.

• **The Combat Innovation & Transformation Cell (CITC):** This cell serves as a translator and connector, aligning incoming proposals to actual Effect needs. It utilizes a four-tier screening model to structure complex information, protecting senior leaders from overload and preserving their time for meaningful choice.

• **Strategic Contracting**: Contracting is viewed as an architectural instrument used to acquire options, speed, leverage, and **sovereignty**. The framework requires tiered Intellectual Property (IP) control (Tier I, II, III).

This ensures the government maintains the ability to modify, sustain, and replicate mission-critical systems through the commercial industrial base or the **Department of Defense Organic Industrial Base (DOIB)** without external dependency.

• **Adaptation and Execution**: The system employs **Centralized Intent** (defined by the CER) with **Decentralized Execution** (performed by units, labs, industry, and the DOIB), enabling flexibility at the edge while maintaining strategic coherence.

C2G asserts that **adaptability, not size, governs survival**, and this continuous evolution is the source of **Competitive Overmatch**. Speed is a moral and strategic imperative, ensuring that timely capability reduces the likelihood of conflict by removing doubt and discouraging aggression. The crux of C2G is to generate combat capability, but its profound by-product is achieving deterrence and resilient peace through ethical, credible, and adaptive strength. The principles of C2G are universal, reflecting the structured adaptation common across all effective organizations and found in natural evolution.

HOW TO READ THIS BOOK

This book introduces an integrated system, not a single idea. Each chapter adds a layer—Effects, governance, industry, contracting, people, and ethics—that together form a single operating model. Many sections overlap in ideas due to the interrelated nature of this system. If any section or concept initially feels abstract, keep going. Every concept is anchored later to how decisions, authority, resources, and accountability actually work.

CHAPTER 1 — DEFINING COMBAT CAPABILITY GENERATION (C2G)

Combat Capability Generation (C2G) is a framework that enables a unified, Effects-driven approach to transforming operational need into rapidly delivered, continuously improving military capability.

Dominance Is Not Enough: The Dinosaur Imperative

Dinosaurs were the uncontested masters of their world. Their scale, strength, and diversity made them the apex of biological capability. Their extinction was not a failure of capability but a failure of adaptability.

C2G is built on the recognition that dominance without agility is temporary. Nations that fail to adapt fast enough eventually face environments they did not anticipate and for which they are unprepared. The purpose of C2G is to ensure that the United States never becomes the strategic equivalent of a dinosaur — powerful, impressive, but overtaken by a world that shifted faster than its institutions could respond.

Adaptability, not size, determines survival. Speed, not mass, governs relevance. Those who adapt — win.

The same pressures that shape survival in the natural world now shape national security. Change is constant. Advantage is fleeting. Systems that cannot adapt at speed do not fail gradually—they are overtaken.

Many of the elements within Combat Capability Generation already exist in isolation; what is new is the deliberate integration, governance, and operational coupling of those elements into a single, adaptive system optimized for speed, adaptation, and scale.

1.1 The Need for a New Way of Thinking

Just as species survive only by adapting to changing environments, military capability survives only by adapting to changing threats, environments, and technologies; the pressures are different, but the imperative is the same.

The modern world is defined by acceleration. Technology evolves at an unprecedented pace. Ideas that once took decades to mature now appear, disrupt, adapt, and vanish within a handful of years — sometimes only months. This rapidity of change has not only transformed commerce and communication, but the very nature of competition, security, and power.

Yet while the world accelerates, many of the systems responsible for national defense continue to move at a measured, deliberate, and often cumbersome pace — anchored in processes that were designed for an earlier era.

C2G arises from this tension.

Previous systems were successful in their time. They safeguarded nations, deterred adversaries, and preserved stability during long periods of global uncertainty. But the conditions that shaped them have changed. Success under those conditions does not guarantee relevance under today's pressures.

C2G is the next evolution.

It connects **need, innovation, governance, ethics, and execution** into a continuous and adaptive cycle.

This supports and defends the constitution of the United States through an effective, nationally aligned construct that creates the most effective military in the most efficient means available. It begins with a simple question:

What Effect is truly needed — and how can it be achieved faster, more responsibly, and more sustainably than any adversary can adapt?

Figure 1.1 illustrates the C2G operating cycle, showing how the system continuously translates a changing environment into actionable, mission-owned capability.

The process begins with the operational environment—defined by conditions, threats, opportunities, and the pace of technological change—which drives the identification of required Effects rather than predetermined solutions. These Effects are captured, prioritized, and analyzed within the Capabilities & Effects Registry (CER), where gaps, dependencies, and overlaps are made visible across the enterprise.

The Combat Innovation & Transformation Cell (CITC) then aligns proposed solutions to those Effects through structured screening, ensuring coherence before decisions are made. Portfolio Acquisition Executives (PAEs), as Effect providers, select pathways and authorize execution through contracting and acquisition mechanisms that preserve modularity, data rights, and adaptability.

Execution occurs across units, industry, laboratories, and the Organic Industrial Base (OIB), with results assessed through operational outcomes and user feedback. This feedback loop continuously informs reassessment and adaptation, allowing the system to evolve at speed while maintaining ethical governance, mission ownership, and strategic control.

Why this matters:
This diagram represents the entire C2G construct in one view. If this operating cycle is understood, the rest of the framework will follow. If it is not, the details that follow will lack context.

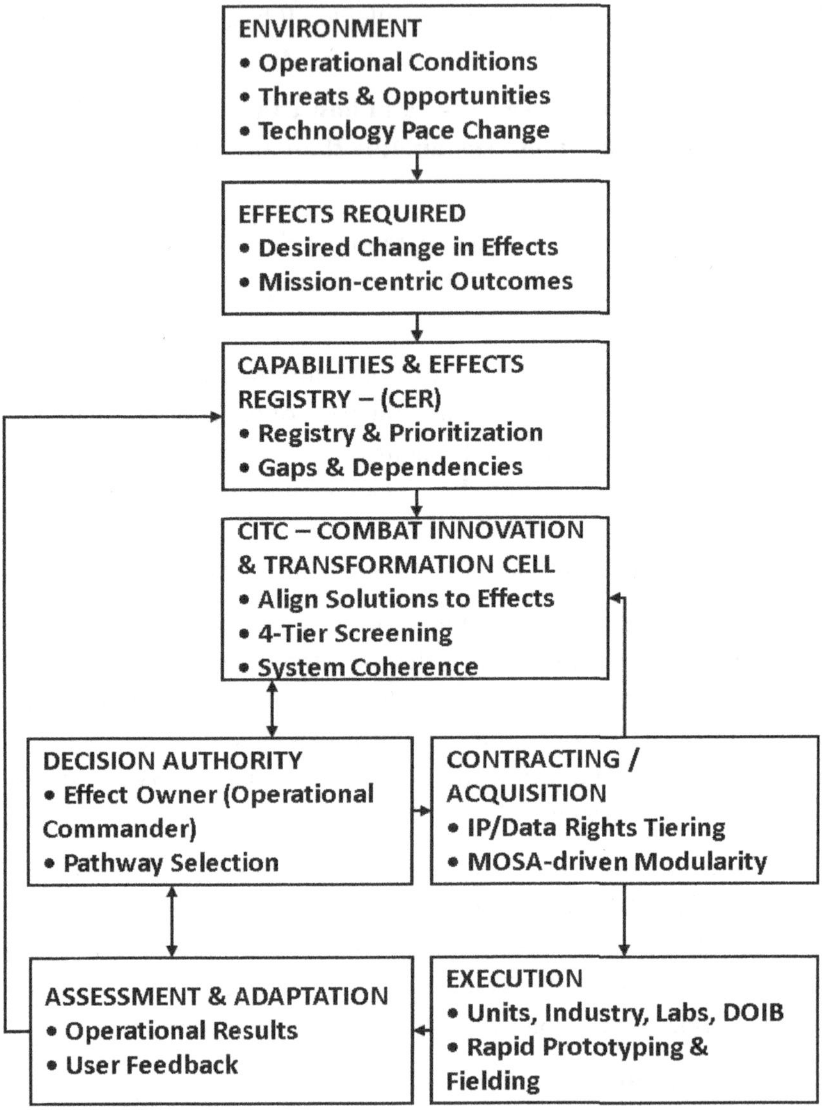

Figure 1.1 – C2G Master Ecosystem Diagram

The Definition of Combat Capability Generation

Combat Capability Generation (C2G) is a framework that enables a unified, Effects-driven approach to transforming operational need into rapidly delivered, continuously improving military capability. In practice, it links operational need to rapid, repeatable, and continuously improving capability delivery through integrated governance, decentralized execution, and disciplined adaptation.

1.2 From Platform-Based Thinking to Effect-Based Thinking

For generations, military strength was measured in platforms: aircraft, ships, ground vehicles, weapons, and systems. Ownership of these assets often defined dominance. This approach made sense in a world where complexity was concentrated in hardware.

The result was a platform-centric ecosystem where innovation, funding, and organizational energy flowed toward ever more sophisticated hardware, rather than toward the operational effects that truly determine battlefield outcomes.

The Army's "Big Five" of the 1980s—Abrams, Bradley, Apache, Black Hawk, and Patriot—perfectly illustrate this era. They were remarkable platforms, but the acquisition system wrapped operational need around hardware, instead of designing hardware around Effects. Complexity accumulated inside individual platforms, not across integrated Effects, which made adaptation slow and modernization cyclical instead of continuous.

Our world has shifted.

The United States now competes in what amounts to a new arms race of adaptation speed. Nations are no longer defined by who fields the largest or most platforms, but by who can sense, decide, and generate capability faster. C2G exists to give the United States the advantage in that race.

Today, power flows through networks, data, integration, adaptability, and decision speed. A platform without adaptability is a liability. A system without ownership of data is a dependency.

C2G does not eliminate platforms. It reframes how they fit within a portfolio to deliver Effects.

In a platform-first model, the platform is the center of gravity. In an Effects-first model, the **mission is the center of gravity.** Platforms become interchangeable contributors rather than monolithic answers. C2G shifts complexity away from hardware and into the ecosystem that generates Effects: sensing, decision-making, integration, adaptation, and feedback.

In C2G, the primary unit of focus is not the item, but the **Effect** it creates:

- Increased survivability
- Accelerated decision-making
- Enhanced awareness
- Reduced vulnerability
- Expanded freedom of action
- Improved availability

Platforms are only valuable in terms of the effects they enable. This subtle shift changes everything.

Instead of asking: "What system should we build?"

C2G asks:

"What change (Effect) is required given a situation, and how do technologies, humans, and structures generate that change?"

This aligns neatly with the Observe and Orient elements of John Boyd's Observe, Orient, Decide, Act (OODA) Loop. It can be applied at all levels from tactical to strategic, by focusing effort. This is where the power of C2G can help to set the conditions and shape the future battlefields – or negotiating tables.

Asking the above question allows multiple solutions to be considered simultaneously and reduces dependence on any single path. It eliminates potential single points of failure while also enabling focus on the Effect.

Adaptation replaces rigidity.

1.3 Speed as a Deterrent

Speed in C2G is not about aggression. Speed is about responsibly demonstrating resolve to win wars as a united force.

A nation that is slow to adapt becomes a target. It creates a power imbalance and a destabilizing force in the world. It creates an opportunity for aggressors and bad actors to seize the initiative. Indecisiveness and weakness invite miscalculation. Delayed capability invites conflict that could have been avoided.

For much of the late twentieth century, the United States operated under a strategic condition often described as the "Pax Americana," where overwhelming military primacy created space and time for long, hardware-centric modernization cycles. Platforms could take decades to field because the United States faced no peer capable of outpacing its development timelines.

That era is over. Strategic competition has collapsed the margin for slow adaptation, and deterrence now depends on the ability to sense change, decide quickly, and generate capability faster than any adversary can exploit. C2G is designed for this environment—it replaces platform-centric thinking with an Effects-driven system built for speed, integration, and continuous adaptation.

C2G recognizes that timely capability reduces the likelihood of conflict by removing doubt, discouraging aggression, and stabilizing balance. Speed strengthens deterrence by compressing an adversary's decision space and signaling that the United States can out-adapt any shift in their strategy.

In this sense, speed is not reckless or impulsive. It is deliberate. It is proactive. It is **ethical.**

This is why C2G is not solely for Soldiers. It also works for the nation and all humanity.

An effective C2G framework creates a powerful deterrent. It eliminates temptation. It lowers the probability of violence.

The faster adaptation occurs, the fewer conflicts are required to prove strength.

Caution: Continuous Improvement and Continuity of Operations

C2G should never reduce our military's effectiveness. Moving and progressing with speed requires measured risk management. While adapting new capabilities, it is imperative to ensure that our current capability is not lost or lowered. In some cases, that may mean building in backward compatibility or even retaining capabilities until the new capability is proven to increase effectiveness in our formations.

Additionally, our military is currently engaged in operations worldwide or training for those operations. This is their mission. During this 'transformation in contact', it is necessary to evolve while the force remains fully engaged. C2G is designed to strengthen the system without pausing it.

1.4 It Is Universal Across All Effective Systems

While C2G is designed to support the military capability generation, it can be applied **to any unit, nation, era, or individual.**

In fact, this concept is visible throughout the natural world. Species survive not because they are the strongest — but because they adapt fastest when their environment changes.

This is evolution.

The organisms that thrive are those that:

- Sense change early
- Respond decisively
- Retain what works
- Abandon what no longer serves
- Evolve continuously

C2G is simply the application of this natural law to the military enterprise. It makes change intentional. It structures adaptation in a high-stakes environment.

This is why its relevance stretches beyond conflict. Its logic applies anywhere complexity and uncertainty exist.

1.5 Humanity as the Foundation of Capability

At the core of every system, every organization, every structure, and every technology is one immutable constant: **people.**

Equipment does not adapt — people do.

Systems do not innovate — people do.

Organizations do not learn — people do.

C2G was not designed to produce machines. It was designed to serve the human purpose — to protect life, preserve stability, reduce the need for conflict, or dominate in combat when needed.

This is why Soldiers are embedded in design environments. This is why civilians are brought into understanding. This is why ethics, legality, and moral responsibility are woven through every phase.

It is not enough to be effective.

C2G must be **right.**

It must align with human values, lawful principles, and long-term sustainability.

Any model that generates power without conscience eventually destroys the very world it seeks to protect.

This work begins with humanity because **without humanity, there is no purpose to defend.**

1.6 The Crux and the By-Product

The purpose of this book is to demonstrate how to **generate combat capability.**

It is not to react to conflict.
It is not to glorify combat.
It is not to prolong war.

C2G is designed to create a clear, adaptive, responsive, ethical capability so that conflict becomes either:

- Decisive and short if forced upon us
- Unnecessary due to deterrence

C2G enables overwhelming combat capabilities to dominate on the battlefield. Through its successful implementation, deterrence can be achieved as a natural **by-product.**

Peace is not created by weakness.
It is created by understanding and credible, adaptable strength.

C2G is the framework by which that strength is built — continuously, ethically, and intelligently.

Chapter 1 Core Truth

C2G is a way of thinking that treats adaptation speed as a strategic necessity rather than a procedural preference.

CHAPTER 2 — FROM REQUIREMENTS TO EFFECTS

Dinosaurs and the Consequences of Failing to Adapt

Nature offers its own lessons in the balance between evolutionary refinement and revolutionary disruption. Dinosaurs dominated the earth for more than 150 million years, not because they were fragile or poorly adapted, but because they were extraordinarily successful within the environments they understood. Their extinction was not the result of inferior capability — it was the consequence of a fundamental mismatch between their highly specialized strengths and a rapidly changing world.

C2G exists to prevent the same mismatch within national defense. Evolutionary improvements strengthen what already works; revolutionary changes redefine what is possible. But both must be understood through the lens of Effects, not platforms. Dominance is temporary when adaptation is slow, and the environment — whether natural or human driven — always changes faster than institutions expect.

This is the central warning for any capability enterprise:
If you cannot evolve as required, or revolutionize, when necessary, the world will move beyond you.

This chapter will discuss the background and purpose of the Effects-based model. The C2G Enterprise View seen in Figure 2.1 demonstrates the people, facilities, and resources responsible for execution. While it is part of the environment, it also has the ability to shape and change the environment around it.

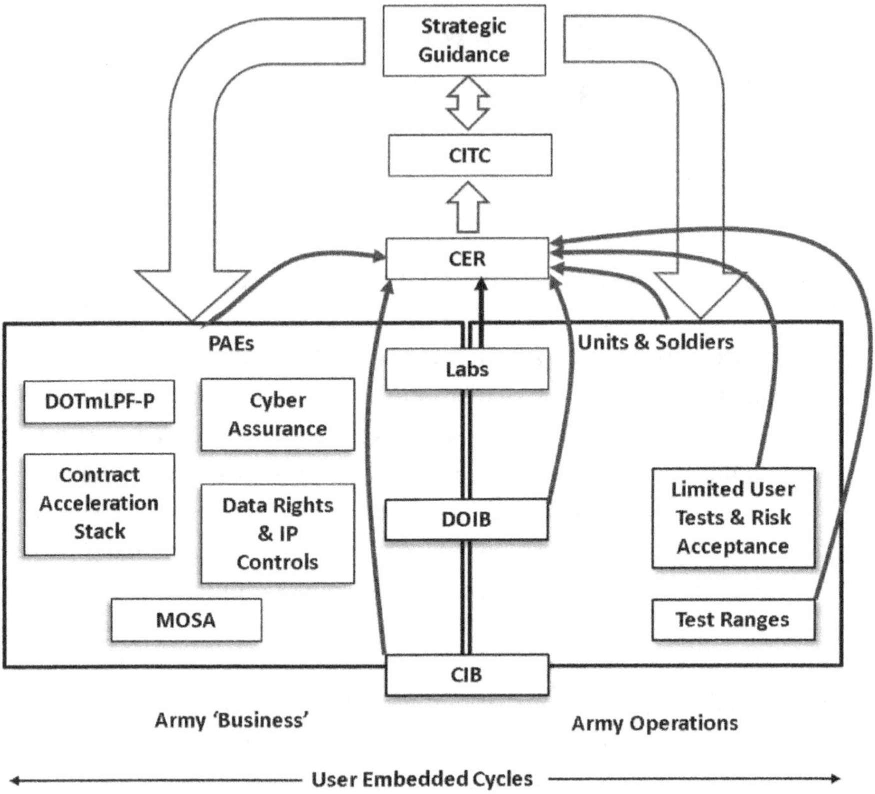

Figure 2.1 – C2G Enterprise View

2.1 The Limits of Traditional Capability Development

For decades, the United States Department of Defense relied upon structured, linear processes to define, develop, fund, acquire, and field military capability. These processes, commonly known through frameworks such as Joint Capabilities Integration and Development System (JCIDS), Planning, Programming, Budgeting, and Execution (PPBE), and the Defense Acquisition System (DAS), were built for an era of relative technological stability and predictable pacing. In that context, long timelines, detailed requirement documents, and multiyear development cycles made sense.

A system that required ten or fifteen years to define and field a new platform could still generate competitive advantage when global innovation moved at an equal, deliberate pace. Industry, academia, and government laboratories operated within the same calendar-based, annually funded rhythm. The battlefield of the twentieth century was defined primarily by hardware, geography, and mass. Major weapons programs set the tone and laid the foundation for what warfare and battles would include.

That world no longer exists.

In the twenty-first century, innovation cycles have drastically compressed. Commercial technology refreshes itself in months, not decades. Software evolves through continuous updates. Digital ecosystems update themselves daily.

Commercial development cycles inform the pace and potential of adaptation, but they do not eliminate the need for military governance, accountability, or sovereign control.

Meanwhile, threats have become decentralized, adaptive, and asymmetric. Under these conditions, a static, checklist-based approach to requirements produces obsolescence before delivery.

Adaptation Speed Example: Obsolescence During Development

An example of the rate of change can be seen during the development of the Army's AH-64E Attack Helicopter. During development of the Version 4 aircraft, a Mission Processor (MP) was included that provided improved capability, integration, and speed compared to the Version 1 aircraft. The Version 4 development was planned to take two years. During that time, the next generation of Multicore Mission Processors (MMP) was released.

The MMP provided exponential processing power compared to the MP which had now essentially become obsolete. To deliver a sustainable product, remain relevant, and provide for adequate growth, the decision was made to simultaneously develop software on both systems. This ensured aircraft could be delivered immediately with the proven MP capability, but would be upgraded to the next generation MMP once they were qualified and produced. This approach also provided for system growth with a synchronized fielding that reduced integration time and training needed.

This example demonstrates adaptation-speed. It shows that capability systems must be able to deliver now while upgrading continuously, without waiting for a perfect end state.

C2G does not discard requirements discipline, budgeting rigor, or program oversight. It actually preserves them. What changes is the organizing principle: decisions are aligned to Effects rather than platforms or artifacts. This shift allows multiple solutions to compete, adapt, or fail without forcing premature commitment — accelerating learning while reducing irreversible error. It aligns progress.

Instead of asking, "What do we need in ten years?" the question becomes:

"What Effect must be achieved now, and how do we adapt as that Effect changes over time?"

The shift from rigid specification to dynamic Effect marks the first and most fundamental departure from legacy thinking.

2.2 Historical Proof: When Speed and Adaptation Changed Everything

History repeatedly demonstrates that decisive advantages arise not merely from superior resources, but from superior adaptation.

The P-51 Mustang did not emerge from a slow, multiyear acquisition cycle. It was born from an urgent wartime need and refined through rapid iteration, cross-national collaboration, and relentless feedback. What began as an experimental airframe evolved into one of the most decisive fighters of World War II.

Even in ground warfare, the ingenuity of Hobart's "Funnies" during the Normandy invasion illustrates how rapid innovation, guided by operational necessity rather than rigid doctrine, can fundamentally change outcomes. Modified tanks, mobile bridges, flail mechanisms, and specialized engineering vehicles transformed seemingly impossible terrain into maneuverable ground.

The Manhattan Project stands as a historical example of what becomes possible when authority, urgency, and adaptation are aligned toward a single, overriding effect. In this case, that effect was unambiguous: end the war. Scientists, military leaders, industrialists, and policymakers were unified under one objective and granted unprecedented authority to act quickly, share information, and adapt dynamically. The Manhattan Project is referenced here to illustrate the relationship between authority, urgency, and adaptation—not as a model for scale, secrecy, or organizational replication.

C2G revisits these authorities and recreates the environment that ushered in unified solutions. It allows adaptation to occur at multiple layers — software, subsystems, and platforms — each operating on different timelines. This accelerates change where it is feasible while preserving deliberate stability where it is required.

Further back in history, long before digital technology, the Industrial Revolution forced armies to adapt to employ trains, engines, mechanized logistics, and mass production. Those who resisted modernizing fell behind. Those who embraced it reshaped warfare entirely.

We are currently at a crossroads where commercially developed technologies and industries, in addition to military development can both be leveraged to increase and improve our current combat generation processes.

In each of the previous historical examples, there remains one unifying factor:

They were not driven by perfect plans. They were driven by urgent effects and rapid adaptation.

C2G does not invent this principle. It codifies it. We must exploit it.

2.3 The Evolution from JCA to CER

The Joint Capabilities Areas (JCA) framework represented a major shift in military planning. It attempted to move the focus away from individual programs and toward broader capability areas such as force application, logistics, protection, and sustainment. In its time, this was visionary.

However, JCA remained a taxonomy. While it mapped the battlefield, JCA did not change with it.

The **Capabilities and Effects Registry (CER)**, within the C2G construct, represents the next evolution.

Where JCA was a static map, the CER is a **dynamic system**. The CER changes in response to an ever-evolving operational environment. It manages required Effects alongside the capabilities to achieve them.

It is continuously updated and prioritized with:

- Emerging threats
- Desired Effects
- Capability gaps
- Candidate solutions
- Operational feedback
- Data rights posture
- Security and supply-chain risk
- Interdependencies between domains

The CER is not a software solution or database construct; it is a governance mechanism supported by information systems.

Figure 2.2 depicts the internal logic and function of the CER, illustrating how required Effects are captured, structured, and managed. The CER serves as the central cognitive mechanism of C2G, organizing mission-owned Effects, associated capability pathways, gaps, dependencies, and risk factors in a single, continuously updated registry.

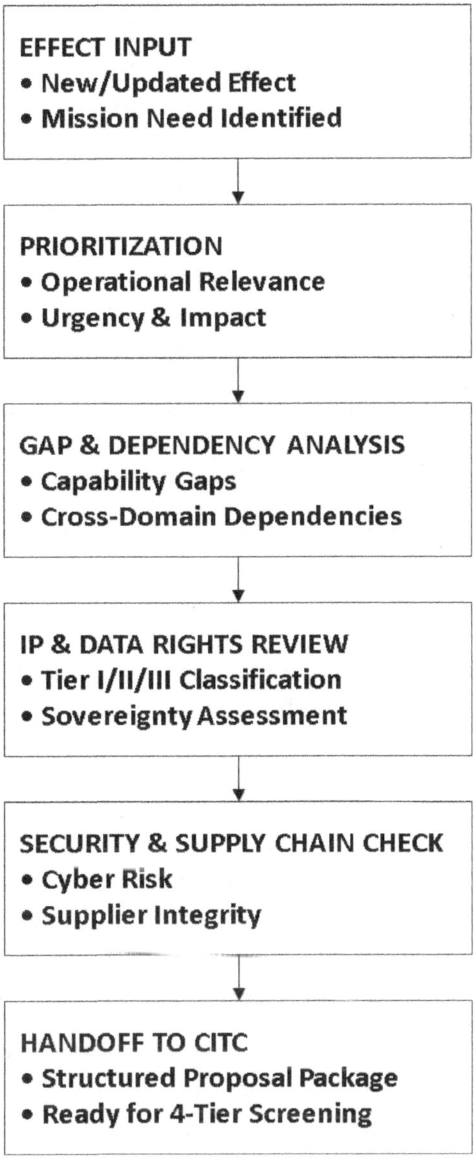

Figure 2.2 – CER Process Diagram

Why this matters:

This is where C2G shifts from philosophy to governance. The CER prevents critical needs from being lost, duplicated, or delayed by making Effects visible, comparable, and actionable.

As operational conditions, threats, or technologies change, Effects are added, reprioritized, or refined, ensuring that the enterprise remains aligned to current reality rather than legacy assumptions. By making parallel efforts, redundancies, and dependencies visible, the CER enables informed decision-making, prevents unintentional duplication, and preserves multiple viable solution paths.

In this role, the CER provides shared situational awareness that allows commanders, portfolio leaders, and governance bodies to exercise disciplined judgment with speed, coherence, and strategic control.

Before examining the internal mechanics of the CER, it is important to understand that it is not a database or IT system, but a governance construct supported by information tools. The following diagram illustrates its logic, not its software implementation.

The CER is not organized by platform or program. It is organized by **Effect**.

This makes the CER the cognitive engine of C2G. It is the mechanism through which leadership, Portfolio Acquisition Executives (PAEs), Capability Portfolio Executives (CPEs), and commanders see reality as it is, not as it was once documented.

Some examples of required Effects could be: “Neutralize Enemy Armor at or before Standoff Range” or “Maintain Continuous Connectivity in Contested Environments”. While these are the Effects required, there may be multiple capabilities to create them. REMINDER: Capabilities can come from any of the DOTmLPF-P domains – not just the material element.

As an Effect enters the CER, it does not begin a slow paperwork journey. Instead, it becomes visible enterprise-wide and immediately open to accelerated solution mapping through the Combat Innovation & Transformation Cell (CITC).

2.4 Dual-Use, COTS, NDI, and GOTS: A Unified Innovation Ecosystem

Modern capability does not come from a single source. It is the product of a **blended ecosystem** made up of:

- Commercial Off-The-Shelf (COTS) technologies
- Non-Developmental Items (NDI) that are modified for military use
- Government Off-The-Shelf (GOTS) solutions

There is also a special category of technology referred to as dual-use. This is a technology that enables a capability that can be used in either civilian/commercial or military applications.

Within the C2G, none of these are competing categories. They are **parallel streams of potential**.

Dual-use is especially important. It operates in two directions:

1. Commercial technologies adapted for military advantage (Innovative application of a technology – Operationally Applied)
2. Military-developed technologies transitioned into civilian applications (Repurposing of a prioritized military Effect into the civilian or commercial sector)

Figure 2.3 illustrates the differing innovation curves across commercial, dual-use, and government-developed systems. C2G does not attempt to force these curves to converge. Instead, it synchronizes them where able, but allows each to contribute at the point where it is strongest. Figure 2.3 should be read as a conceptual comparison of innovation tempo rather than a literal timeline.

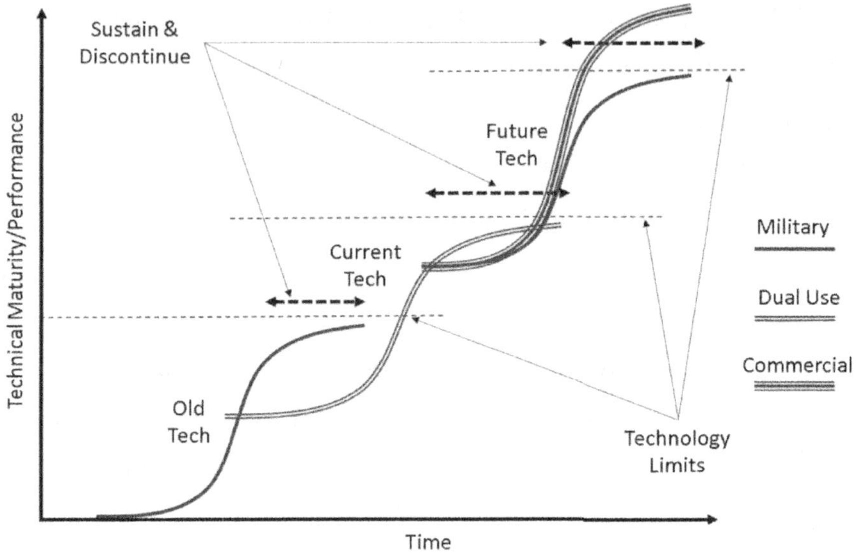

Figure 2.3 – Innovation Curves

These movements are not accidental. In C2G, they become strategic.

The National Industrial Base benefits from military innovation, just as the military benefits from commercial speed and creativity. Under the C2G model, leveraging COTS and commercial solutions benefits both the commercial sector for stability and military operations for overmatch.

Instead, it is part of **mobilizing and strengthening the National Industrial Base as a whole.** This reinforces the Economic instrument of national power.

Within this system, the Department of Defense Organic Industrial Base (DOIB) exists to **augment and continue** where commercial capability becomes fragile, restricted, compromised, or unavailable. This relationship is symbiotic, not adversarial.

From the beginning, each solution — regardless of origin — is evaluated not only on performance, but also on:

- Cybersecurity posture
- Supply-chain integrity
- Export limitations
- Intellectual property rights
- Government control and continuity
- Effectiveness in a given situation

A solution that performs brilliantly but places the nation in dependency is not a strategic solution.

With C2G, **sovereignty is a designed feature, not an afterthought.**

Modular Open Systems Approach (MOSA) as a Cross-Sector Integrator

MOSA bridges military, commercial, and dual-use ecosystems by enabling open interfaces that allow technologies from any origin — COTS, dual-use, NDI, or GOTS — to integrate rapidly and securely. This eliminates artificial boundaries between sectors and accelerates effect delivery while maintaining sovereignty and cybersecurity.

2.5 Evolutionary and Revolutionary Change in Capability Development

While MOSA enables rapid integration, it generally *enables* evolutionary, incremental growth. As an approach, and more so as an architecture, MOSA requires fundamental rules, formats, and standards to be followed. Revolutionary changes by their very nature frequently do not follow traditional rules, formats, or standards.

Technological progress occurs along two distinct but interconnected paths: **evolutionary change** and **revolutionary change**.

Evolutionary change occurs through incremental improvement — the familiar, steady refinement of systems such as vehicles, radios, protective equipment, or power systems. These upgrades improve performance but do not fundamentally change the underlying paradigm. Evolutionary modifications generally require modest adjustments across DOTmLPF-P and present fewer integration risks because the surrounding system remains stable.

Revolutionary change is different. It disrupts an existing paradigm entirely.

The introduction of the smartphone is one such example: a single device integrating touchscreens, sensors, software ecosystems, and connectivity in a way that rendered entire categories of legacy technologies obsolete.

Revolutionary changes deliver new pathways to achieve effects, but they require **greater institutional investment** to realign doctrine, training, sustainment, facilities, leadership expectations, and policy. Both forms of change are essential. Evolution strengthens and refines. Revolution reshapes what is possible.

An Effect-based view helps clarify the relationship between evolutionary and revolutionary change by separating what must be preserved from what can change. Figure 2.4 uses the evolution of music media as an illustrative example, showing how fundamentally different technologies—some evolutionary, some revolutionary—can coexist across overlapping curves while delivering the same underlying effect. This analogy is intentionally non-military to emphasize that Effects persist even as delivery mechanisms change.

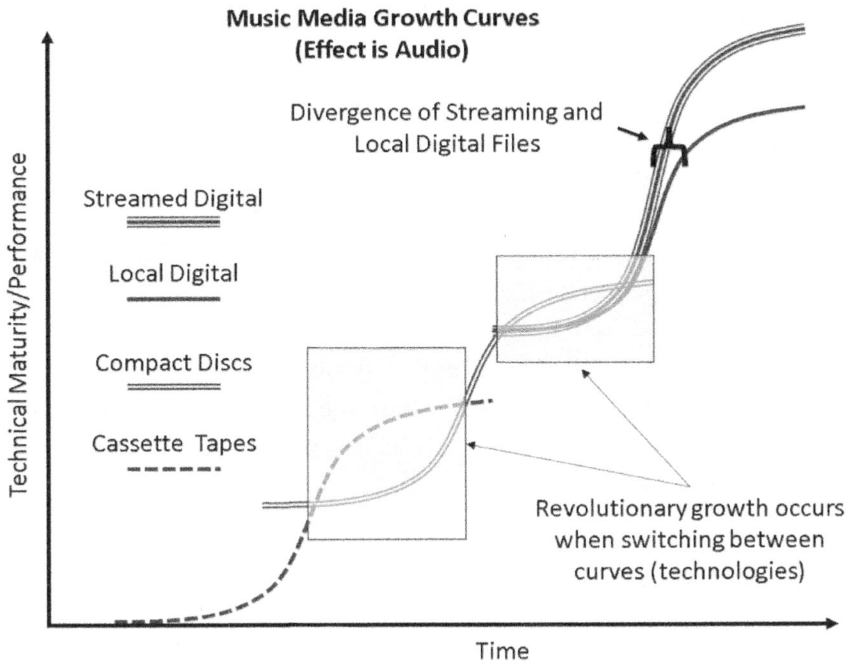

Figure 2.4 – Music Media Growth Curve Example

An Effect-based view helps make sense evolutionary and revolutionary changes. The music industry illustrates these principles: vinyl, 8-track, cassette, compact disc, and streaming are entirely different technologies — some evolutionary, some revolutionary — yet they all serve the same effect: **delivering music to the listener**.

Revolutionary shifts changed how the Effect was achieved; evolutionary steps improved reliability, portability, and cost. C2G embraces both, ensuring we never confuse the method for the purpose. The same pattern applies to military capability.

2.5a Second- and Third-Order Considerations

While all changes require a holistic review of the entire DOTmLPF-P domain (further explored in Chapter 7), revolutionary advances require a more deliberate analysis of second- and third-order consequences.

An innovative breakthrough that accelerates capability delivery may also create new dependencies, expose supply-chain vulnerabilities, increase training demands, or shift adversary adaptation patterns. Evolutionary change generally carries lower systemic risk, while revolutionary shifts require institutional awareness of the broader ecosystem they disrupt. C2G recognizes that innovation without foresight is merely acceleration toward an unknown outcome.

2.6 From Requirements to Effects

Traditional acquisition asks for a list of performance characteristics such as Speed, Range, Size, Capacity, and Endurance.

C2G asks a different, more fundamental question:

"What change (Effect) do we need in the operational environment?"

Figure 2.5 illustrates the Effects Taxonomy used within C2G. Rather than organizing capability around platforms, programs, or technical specifications, this taxonomy organizes thinking around the changes we seek to produce in the operational environment. Each category represents a class of effect that can influence mission outcomes independent of any single system or configuration.

Why this matters:
This taxonomy replaces platform-centric requirements with outcome-centric thinking. It gives leaders a common language for deciding what change in the operational environment truly matters.

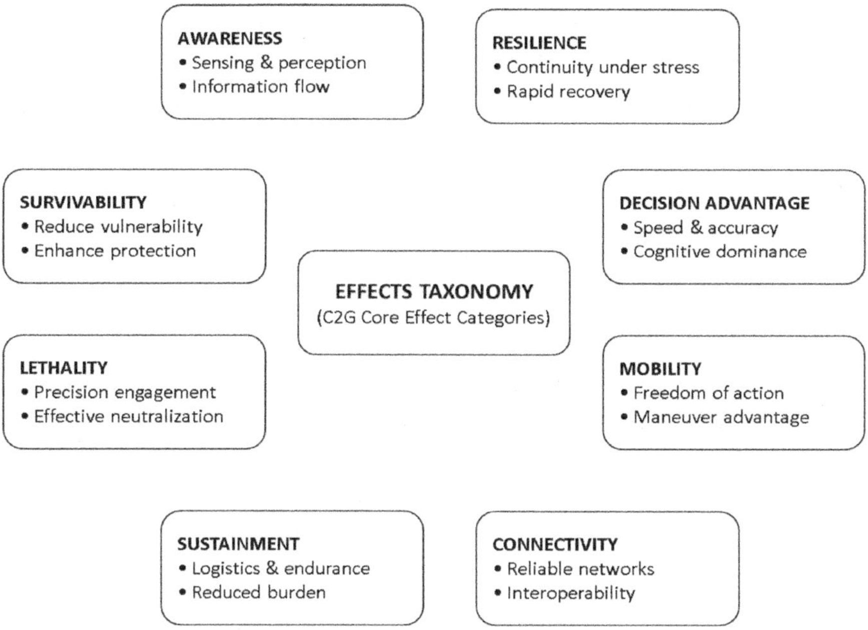

Figure 2.5 – Effects Taxonomy

The taxonomy is intentionally non-prescriptive. It does not dictate solutions, architectures, or technologies. Instead, it provides a shared framework for identifying, discussing, and prioritizing effects across operational, technical, and institutional boundaries. Multiple capabilities—delivered through different means, at different speeds, and at different levels of maturity—may contribute to the same effect.

By grounding discussion in Effects rather than requirements, this taxonomy enables adaptability, competition, and innovation. It allows decision-makers to evaluate alternatives based on contribution to outcome rather than conformity to specification, and it creates space for parallel solutions that can be fielded, scaled, or retired as conditions evolve.

This taxonomy serves as the foundation for the CER and informs how the CITC aligns emerging solutions to operational need. It is not a checklist—it is a lens through which capability is understood, compared, and generated.

By focusing on Effect instead of configuration, the system becomes adaptive rather than brittle. Multiple solutions can be mapped to the same Effect but provided through alternate, complimentary means. This encourages competition, redundancy, creativity, and speed.

To summarize, the CER captures this logic, and the CITC aligns innovation to meet it.

In this model, the warfighter does not generate or receive a list of technical specifications. They receive an **outcome.**

2.7 Redundancy Versus Duplication: An Intelligent Strategy

C2G deliberately distinguishes between **wasteful duplication** and **strategic redundancy.**

Redundancy, when intentional and visible, is a source of resilience; duplication without awareness is waste.

Why this matters:
This distinction determines whether redundancy produces resilience or waste. C2G is designed to allow multiple pathways to success without tolerating uncontrolled overlap.

INTENTIONAL REDUNDANCY (GOOD)	UNINTENTIONAL DUPLICATION (BAD)
✓ Parallel solution pathways	x Uncoordinated efforts
✓Competition increases speed	x Waste of resources
✓ Diverse approaches reduce risk	x Creates conflicting systems
✓ Strengthens adaptability	x Slows modernization

C2G DISTINGUISHES BETWEEN:
Redundancy = resilience & speed
Duplication = inefficiency & friction

Figure 2.6 – Redundancy vs. Duplication

History offers many examples of parallel development paths: Beta vs VHS, multiple engine designs, competing aircraft platforms, simultaneous research in artificial intelligence, propulsion, and communications.

While one path may ultimately dominate and become a market leader, redundancy ensures that:

- Failure of one line does not equal total failure
- Innovation pressure remains high
- The best solution emerges organically
- The adversary cannot predict a single collapse point

This philosophy aligns directly with Middle Tier Acquisition (MTA) and Other Transaction Authority (OTA) models, where multiple prototypes are encouraged rather than suppressed.

Redundancy is a **hedge against uncertainty.**

Duplication without awareness is waste.

Redundancy with visibility is strategy.

Through the CER, leaders can observe parallel efforts and ensure they are intentional rather than accidental. SABOT rounds and Hellfire Missiles (both armor destroying Effects) are a great example of a deliberate redundancy to yield a common battlefield Effect.

2.8 Intellectual Property and Data Rights as Strategic Terrain

In C2G, intellectual property is not a legal footnote. It becomes **operational terrain** that is necessary to accelerate the mobilization of the enterprise.

A nation does not truly possess a capability if it cannot:

- Modify it
- Maintain it
- Replicate it
- Transfer it
- Train it
- Defend it from compromise

Within the C2G framework, every capability is assessed across five interconnected dimensions:

1. Effectiveness
2. Speed
3. Security
4. Scalability
5. **Control**

Control is defined by **intellectual property access, data rights, and manufacturing authority.**

Therefore, C2G applies a **tiered, risk-based data-rights model**, governed by mission importance.

Why this matters:
Control of data and design determines who truly owns a capability. Without it, speed, resilience, and sovereignty are illusions.

TIERED DATA RIGHTS MODEL (TIER I / TIER II / TIER III)
TIER I — UNLIMITED RIGHTS • Full government access & reuse • Highest sovereignty • Ideal for MOSA & long-term sustainment
TIER II — GOVERNMENT PURPOSE RIGHTS • Shared access with restrictions • Reuse for government missions only • Moderate sovereignty
TIER III — LIMITED / RESTRICTED RIGHTS • Vendor-controlled • No reuse without permission • Lowest sovereignty; POI risk

Figure 2.7 – Tiered Data Rights Model

Tier I — Mission-Critical / Strategic Systems

For these systems, the government must require **Unlimited Rights or Government Purpose Rights without exception**. This includes:

- Full technical data
- Access to source code where relevant
- Manufacturing transfer authority to the DOIB
- Sustained modification and repair rights
- Allied-transfer permissions as needed

These systems cannot be subject to commercial denial of access. Dependency is unacceptable.

Tier II — Operationally Important Systems

These systems allow negotiated rights, but must include:

- Government sustainment access
- Source code or technical data escrow
- DOIB continuation clauses
- Licensing rights adequate for independent modification

Tier III — Peripheral / Support Systems

These may remain under commercial structures but must still meet:

- Cybersecurity standards
- Supply-chain transparency
- Conflict-of-interest screening
- Interoperability requirements

Under C2G, **a solution that restricts sovereign control is automatically downgraded until mitigated** — regardless of performance.

This is not anti-industry.

It is **pro-responsibility**.

Industry remains a partner — but sovereignty is never surrendered.

Chapter 2 Core Truth

Sustainable capability advantage comes from organizing decisions around mission-owned Effects rather than platform-specific requirements.

CHAPTER 3 — INTELLIGENCE OF SYSTEMS: MDMP, CER, AND CITC

3.1 The Shift from Linear Planning to Adaptive Reasoning

Traditional military planning and acquisition processes have largely followed linear sequences: define a requirement, develop a solution, acquire a system, field a capability, modernize, and replace. These models assume predictable timelines, fixed threats, and relatively static environments.

That assumption no longer holds.

The modern environment is not linear. It is dynamic, ambiguous, and increasingly shaped by the velocity of information rather than physical production alone. In such an environment, static planning becomes a vulnerability.

C2G recognizes that intelligence must now exist not only in the battlefield sense — but in the **system itself**.

In C2G, intelligence is embedded in:

- How needs are identified
- How options are explored
- How solutions are filtered
- How Effects are aligned
- How decisions are justified

This is not simply faster acquisition. It is **smarter structure**.

3.2 The Military Decision-Making Process Reimagined

The Military Decision-Making Process (MDMP) is designed to provide commanders and staffs with a structured approach to complex problem solving. It emphasizes understanding the environment, analyzing options, evaluating risk, and selecting a course of action based on logic rather than emotion or assumption. It has been very successful when applied appropriately.

These principles are not confined to the battlefield.

C2G adopts the logic of MDMP as a **cognitive framework** for capability development, without replicating tactical planning procedures:

1. Understand the environment
2. Define the true problem
3. Identify constraints
4. Explore multiple options
5. Evaluate second- and third-order effects
6. Assess risk
7. Select the most logical path

This same mental discipline is applied when:

- Interpreting operational gaps
- Assessing proposed solutions
- Evaluating emerging technologies
- Aligning solutions to Effects
- Prioritizing limited resources

In both contexts — operational and developmental — the best decision is the one grounded in structured logic, not emotion, attachment, or tradition.

This alignment between battlefield cognition and capability generation creates a **shared intellectual language** across the force.

It also allows both senior leaders and junior Soldiers to contribute meaningfully, using a skillset already embedded in military culture.

Case Study: XM-25

The U.S. Army was developing a revolutionary new weapon called the XM-25. This system was designed to deliver an operational capability but had faced several technological development challenges. As prototypes were developed and tested, the decision was made to deploy it to the war in Afghanistan to gain operational insight and provide a new capability to Soldiers.

During this limited, operational experiment, misunderstandings arose between some of the operational forces and the development team. The failure to include operational Soldiers during the development combined with limited, in-theater training, and technical issues had created an emotionally driven, unfavorable, and divided opinion on this weapon. The technology 'Valley of Death' was widened.

On the operator side, some viewed it as heavy, complex, and a disruption to current tactics and doctrine. Others viewed it as a radical shift at the tactical level that would create dominating overmatch at the squad level – IF employed appropriately. The opponents allowed emotion and tradition to cloud their judgement. This sentiment carried out through command channels and into the Pentagon. As a result, the decision was made to terminate the program.

This example illustrates the complexities of governance, adaptation, inclusion, and decision-making dynamics, rather than a claim of direct equivalence across all acquisition programs.

3.3 The Capability and Effects Registry (CER)

At the heart of C2G lies the **Capability and Effects Registry (CER)**.

The CER is not a database of equipment. It is a living registry of **desired Effects**.

The following diagram illustrates how Effects move from operational need to capability delivery, showing the relationship between the

Commander, CER, CITC, Contracting, Execution, and Assessment. This governance structure ensures decisions are mission-owned, innovation is aligned to Effects, and adaptation is continuous.

Figure 3.1 should be read from the top left and follow the arrows as a flow of authority and feedback, not as an organizational chart.

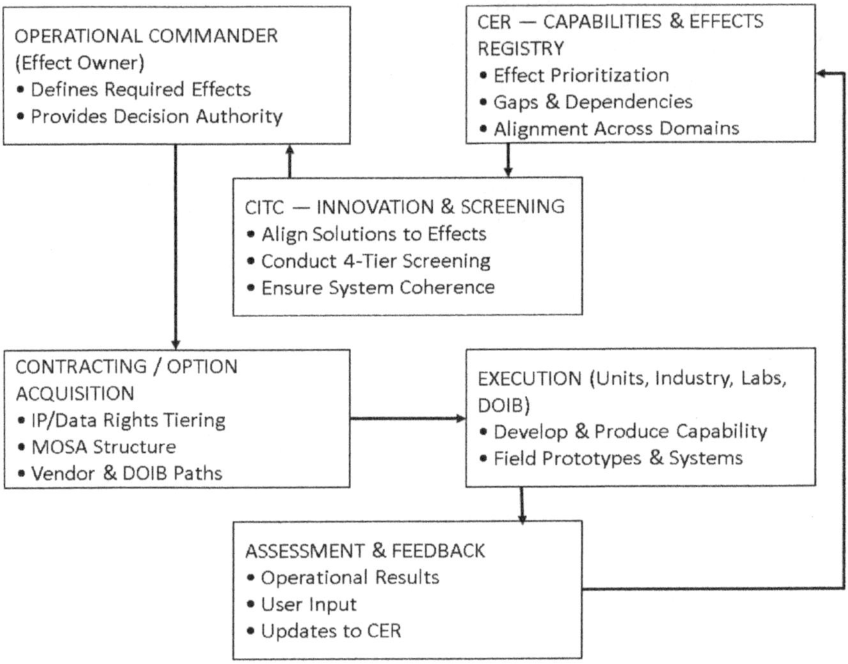

Figure 3.1 – Governance Flow Diagram

Why this matters:

This diagram shows how insight, authority, innovation, and execution remain connected without becoming centralized. It is the governance engine that allows speed without chaos.

Each entry in the CER represents an articulated outcome such as:

- Reduction of vulnerability in a certain domain
- Increase in survivability under specific conditions
- Enhancement of awareness, responsiveness, or resilience
- Improvement in sustainability, lethality, or deterrence posture

The CER answers one fundamental question:

What change must occur in the operational reality to create advantage?

Entries in the CER are not owned by branches or communities. They are owned by **mission relevance**.

Anti-armor is not an aviation problem.
Anti-armor is an Effect.

Mobility is not a ground problem.
Mobility is an Effect.

Resilience is not a cyber problem.
Resilience is an Effect.

This prevents competition between communities and instead encourages collaboration around outcomes.

The CER belongs, appropriately, at the **strategic coordination level**, ensuring alignment across portfolios while preventing duplication of effort.

3.4 The Combat Innovation & Transformation Cell (CITC)

Where the CER defines the **"what"**, the **Combat Innovation & Transformation Cell (CITC)** engages with the **"how."**

The CITC is not a gatekeeper whose primary role is to deny ideas.
It is **a translator and connector**.

Its responsibility is to:

- Interpret Effect requirements
- Filter incoming ideas and proposals
- Align submissions to actual needs
- Connect innovators with stakeholders
- Maintain system coherence
- Prevent overload of decision authorities

The CITC **does not decide** what is funded.
It **enables clear decision-making** by ensuring coherence.

This transforms relationships from adversarial ("approve or deny") to collaborative ("clarify and align").

In this way, the CITC becomes a **capacity amplifier**, not a bureaucratic barrier.

3.5 Preventing Duplication of Effort

One of the most corrosive inefficiencies within large organizations is unintentional duplication: multiple entities unknowingly working toward the same outcome through disconnected paths.

In C2G, duplication is prevented through alignment at the **Effect level**, not the program level.

The CER:

- Tracks what Effects are being pursued
- Identifies overlapping areas
- Allows multiple approaches with redundant purpose
- Preserves diversity of solution while preventing waste

This enables a critical distinction:

Redundancy of capability may be beneficial; duplication of effort is not.

Intentional parallel investment = strength
Unintentional overlap = waste

3.6 The Four-Tier Screening Model

To prevent overload and maintain relevance as submissions increase, C2G incorporates a four-tier filtering system:

Tier One — Relevance Screening

- Does this align to a registered Effect?
- If no, it is deferred, not discarded.

Tier Two — Feasibility Screening

- Is this technically achievable in a meaningful timeframe?
- Can it be matured or scaled?

Tier Three — Risk & Dependency Analysis

- Would adoption introduce dangerous reliance?
- What are the supply chain and cyber implications?

Tier Four — Priority Alignment

- How critical is this Effect right now?
- What is the opportunity cost?

Why this matters:
This model protects senior leaders from overload. Without disciplined filtering, speed collapses under volume and decision-making degrades.

TIER 1 — RELEVANCE
- Aligns to Registered Effect
- Mission Need Validated

TIER 2 — FEASIBILITY
- Technically Achievable
- Maturity & Scalability

TIER 3 — RISK & DEPENDENCY
- Cyber & Supply Chain Risk
- Sovereignty & IP Assessment

TIER 4 — PRIORITY ALIGNMENT
- Operational Urgency
- Resource Fit & Opportunity Cost

Figure 3.2 – CITC Four-Tier Screening Model

Only concepts that pass through all tiers reach senior decision-makers, preserving their time for meaningful decisions rather than volume management.

This maintains **speed without saturation.**

3.7 Logic Over Emotion in Selection

C2G requires the discipline to resist emotional attachment to familiar platforms, historical solutions, or charismatic ideas.

A solution that feels right may not be right.

Therefore, every proposed pathway must justify itself through:

- Alignment to Effect
- Evidence of viability
- Risk profile
- Sustainability
- Contribution to resilience

This does not remove intuition.
It ensures that intuition is supported by analysis.

This intellectual rigor is essential to the credibility and effectiveness of the model.

3.8 Effects Are Mission-Owned, Not Domain-Owned

In legacy systems, ownership of mission space often defined ownership of resources:

- Aviation controlled air power
- Armor controlled ground power
- Signals controlled information

In C2G, this thinking is replaced by **Effect ownership** rather than **domain ownership**.

Since Effects are operational outcomes, not technical artifacts, they therefore must be owned by the commanders responsible for achieving those outcomes in the battlespace. Under C2G, the **operational owner of each Effect is the Joint Force Component Commander whose mission set the Effect supports**—whether land, maritime, air, cyber, space, or special operations. Combatant Commanders provide strategic context, while Service Chiefs provide forces and capabilities under Title 10, but the ownership of an Effect rests with the commander directly accountable for generating that outcome in real operations.

The CER registers and structures Effects, and the CITC analyzes pathways to achieve them, but **neither owns the Effect**. Ownership remains firmly with the operational leader who must deliver that Effect against an adversary.

If the Effect is anti-armor, the solutions may involve aviation, ground forces, robotics, sensors, cyber, or a hybrid of all.

This prevents any single organization from inadvertently extinguishing innovation by claiming or rejecting it on domain grounds.

The owner of an Effect is not a branch.

It is the **mission itself**.

Chapter 3 Core Truth

Speed without structured reasoning creates chaos, while intelligence embedded in governance enables decisive adaptation.

CHAPTER 4 — THE HUMAN ELEMENT

The nature of humanity is ultimately life. However, the nature of war is human. The entire C2G model is designed for the protection of human life, but also it requires complete integration of specially trained and equipped humans to succeed.

4.1 Soldiers as Contributors, Not Just End Users

Every discussion of capability tends to drift toward technology: sensors, platforms, algorithms, systems, and networks. These components matter.

But no capability, no matter how advanced, exists independently of the humans who employ it.

Too often, individuals are positioned solely as *operators* of tools developed elsewhere, by unseen stakeholders and distant institutions. In this type of model, Soldiers become recipients of capability rather than contributors to its creation.

C2G rejects this reduction.

In C2G, **people are not recipients of capability — they are its origin. Users are the bottom line!**

The model recognizes that the most valuable insight comes from those experiencing friction, delay, limitation, and danger firsthand.

Our servicemembers are not simply end users. They are **primary sources of truth**.

Therefore, their role must expand from *operator* to *participant* in the generation of capability.

4.2 Embedding, Not Merely Integration

Traditional approaches often speak of **integration**: integrating users into training cycles, integrating operators into testing, integrating feedback into requirements. While well-intentioned, integration implies something external being added to a pre-existing structure.

C2G requires something more intentional and intimate.

It requires **embedding**.

Embedding is participatory, not advisory.
It is immersive, not observational.
It is continuous, not episodic.

Figure 4.1 shows the entire operating loop into which personnel must be embedded.

Why this matters:
This is where humans are embedded into the system. C2G only works if the people who bear operational risk are part of the process that shapes outcomes.

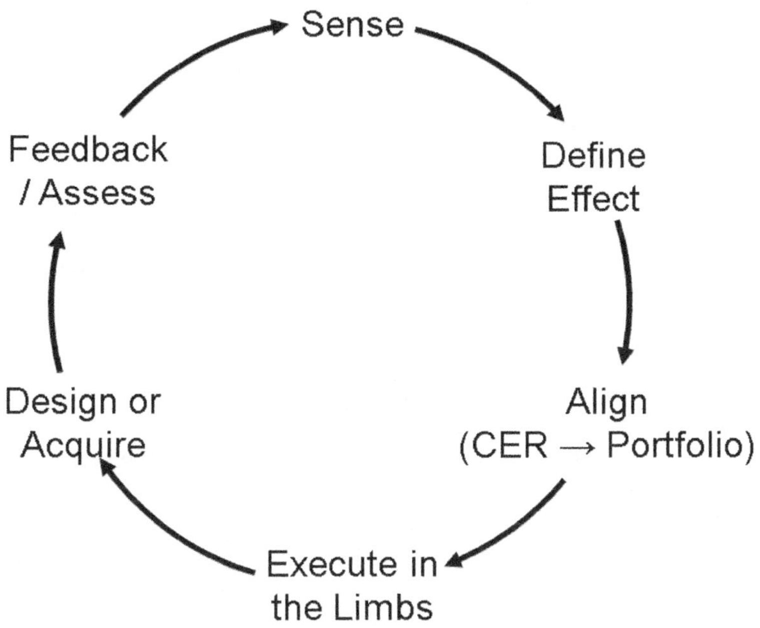

Figure 4.1 – C2G Operating Loop: The Adaptive Cycle

Embedded personnel must participate from inside:

- Innovation cells
- Prototyping teams
- Industrial design environments
- Requirements interpretation groups
- Testing and evaluation cohorts

They are present not to supervise, but to **inform reality** — to bring the lived experience of operations into environments that are often insulated from it.

This provides a dual benefit:

1. Capability developers receive unfiltered insight into reality.
2. Soldiers gain exposure to the larger system that supports them.

This mutual awareness strengthens both.

Embedded human presence becomes a **bridge between theory and truth.**

4.3 The Critical Role of Junior Soldiers

It is tempting to assume that only senior leaders possess the perspective required to shape future capability. In reality, the opposite is often true.

Junior Soldiers stand closer to friction.
They encounter barriers more frequently.
They adapt on instinct.
They see inefficiency long before it appears in reports.

The point of greatest vulnerability is also the point of greatest awareness.

C2G therefore creates deliberate mechanisms to include junior personnel in limited, protected, and structured ways, such as:

- Anonymous capability submissions
- Controlled focus groups
- Embedded observation rotations
- Prototype-user exchanges
- Data-capture environments

This is not about granting disproportionate authority.

It is about **granting voice to innovative forces.**

Their experiences form raw data that informs higher-level reasoning, allowing empathy to convert into design standards.

In a world where senior leaders often see abstraction, junior Soldiers see reality. C2G honors that perspective without compromising command authority.

4.4 Required Education, Training, and Experience

Selection for embedded roles within the C2G ecosystem must balance accessibility with responsibility.

These positions require individuals who possess:

- Professional maturity
- Ethical clarity
- Intellectual curiosity
- Emotional discipline
- Systems-level thinking
- Respect for confidentiality
- Clear communication skills

Ideal candidates may possess combinations of:

- Operational experience
- Technical training
- Engineering exposure
- Systems analysis education
- Acquisition familiarity
- Cyber awareness
- Strong writing and analytic capability

However, the C2G model is intentionally flexible.

It does not assume that brilliance resides only by rank or academic pedigree. Instead, it looks for **insight, adaptability, and character.**

This ensures that C2G remains inclusive without becoming diluted.

4.5 Civil–Military Understanding and Mutual Benefit

The divide between military and civilian communities is often exaggerated by misunderstanding on both sides.

Many civilians view the military as distant, rigid, or violent. Many Soldiers view civilian life as disconnected or naive.

This divide weakens the nation.

Embedding service members into innovation hubs, civilian organizations, educational institutions, and industrial environments builds understanding in both directions.

Civilians gain the opportunity to see:

- The discipline behind the profession
- The humanity behind the uniform
- The ethics behind service
- The responsibility behind authority

Soldiers gain:

- Exposure to diverse thought
- Understanding of social expectations
- Experience in collaborative problem-solving
- Reduced cultural shock upon transition

This does not diminish the profession of arms.
It **enriches it.**

It also strengthens the legitimacy of the system in the eyes of the people it exists to protect.

4.6 Reducing Reintegration Shock

One of the most challenging aspects of military service is not the entrance — it is the exit.

Veterans often struggle not because they lack ability, but because they are disconnected from the world they once defended. Additionally, industry may not have an appropriate understanding of the true abilities of our Veterans.

By embedding Soldiers in hybrid environments throughout their careers, the C2G model:

- Normalizes interaction with civilian systems
- Reduces psychological barriers to transition
- Expands professional vocabulary
- Builds transferable skills
- Preserves dignity in identity shift

The result is not a weakened force — but a **healthier society**.

Service becomes more visible and less isolating. Reintegration to the civilian sector becomes less traumatic.

C2G does not end when someone leaves the uniform.

Its influence continues into the society they rejoin.

4.7 Humanity as Strategic Infrastructure

Within every system, there are visible and invisible components.

The visible includes platforms, networks, facilities, and processes. The invisible includes trust, reason, empathy, and conscience.

C2G is sustained not by the visible, but by the invisible.

Its success depends on the integrity, clarity, and compassion of those who shape it.

This chapter exists to make one truth permanent within this book:

Technology is a tool. Humans are the infrastructure. Humanity is the purpose.

Any model that forgets this truth will, over time, collapse beneath its own power.

C2G does not exist to dominate others.

It exists to preserve stability, deter aggression, protect life, and reduce the need for conflict in an increasingly volatile world. It exists to serve humanity. And therefore, humanity must be embedded — not only in process, but in spirit — at every stage.

Chapter 4 Core Truth

No capability system can adapt faster than the humans empowered to shape, test, and learn within it.

CHAPTER 5 — MOBILIZING THE INDUSTRIAL BASE

5.1 The Industrial Base as a Strategic Weapon System

In traditional doctrine, the industrial base was considered a supplier of warfighting systems.
In C2G, the industrial base is a **component of the weapon system itself**.

History underscores this truth. The side that can build, adapt, repair, and scale faster than its adversary gains a decisive advantage without having to fire a single round. Industrial capacity is not only about production — it is about **option creation**.

Under C2G, the industrial base is reframed as a living system composed of:

- The Commercial Industrial Base (CIB)
- The Department of Defense Organic Industrial Base (DOIB)
- Research institutions and laboratories
- Small innovative firms and non-traditional contractors
- Academia and federally funded research centers
- Local manufacturing networks and artisans

These are not separate actors operating in competition.

They form a **unified capability ecosystem** whose purpose is to translate human creativity into operational relevance.

In C2G, industrial capacity is not measured only in factories and output. It is measured in **adaptability, responsiveness, survivability, and independence.**

The industrial base becomes a mirror of the battlefield:

- Decentralized
- Redundant
- Adaptable
- Resilient
- Interconnected
- Difficult to disrupt

This is not bureaucracy.

This is strategic design.

5.2 The Commercial Industrial Base: Speed, Scale, and Creativity

The commercial sector has mastered speed.

Driven by competition, market pressure, and consumer demand, it innovates at a pace unmatched by traditional defense programs. From microelectronics to software, communications to robotics, the private sector continually compresses development cycles.

C2G does not attempt to outpace commercial innovation.

It **harnesses it.**

The commercial industrial base provides:

- Rapid prototyping
- Agile software development
- Cost efficiency through scale
- Access to global innovation networks
- Continuous iteration
- Market-tested user experience

Commercial firms are often less encumbered by internal policy, giving them the flexibility to test, fail, pivot, and evolve at speed.

In the C2G construct, the role of the commercial base is to act as a **primary innovation engine**, offering:

- Novel approaches to enduring problems
- Unconventional solutions unburdened by tradition
- Technologies developed for civilian use but applicable to operational needs
- Modular, scalable, and upgradeable components

This does not mean the military abandons its technical expertise.

It means it learns to **collaborate at the speed of relevance.**

"Who dares wins" in this context is about decisive integration of ideas.

The following sections contain figures that translate this conceptual relationship into a structural view of the national and organic industrial base.

5.3 The DOIB as an Alternate Innovation Engine

Historically, the DOIB was viewed primarily as a sustainment entity including depots, arsenals, shipyards, and maintenance facilities. That perception is incomplete when viewed in C2G.

The DOIB becomes **an alternate innovation engine, not a secondary one.** Figure 5.1 demonstrates the collaborative, complementary, and competitive nature of the entire Organic Industrial Base.

Why this matters:
Industrial capacity is not a supporting function—it is a strategic weapon system. The ability to build, adapt, and scale faster than an adversary creates advantage before conflict begins.

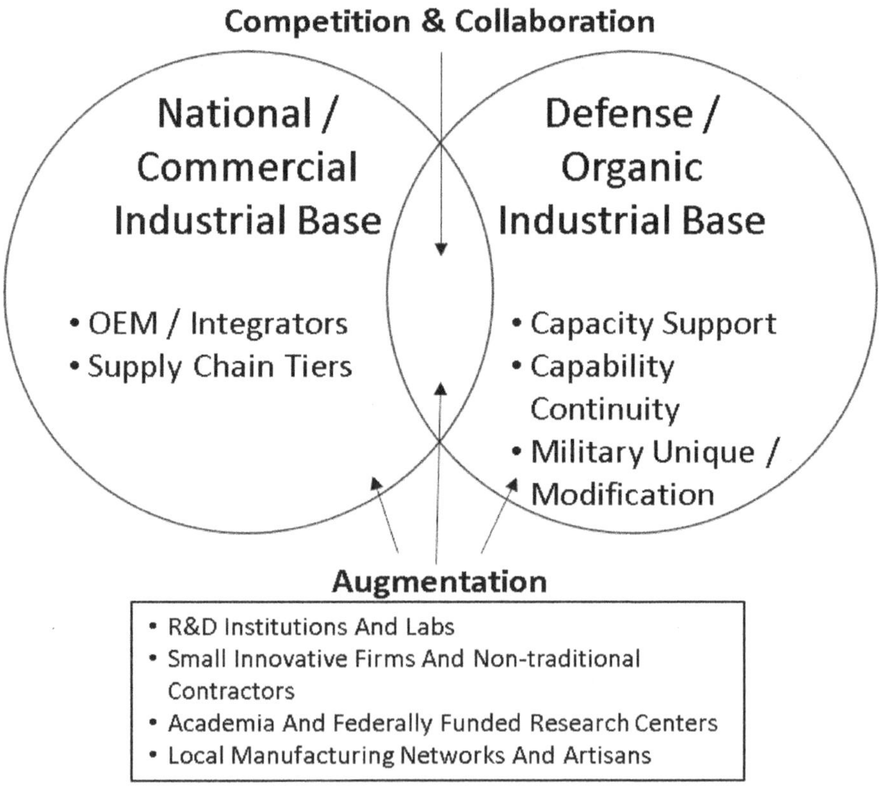

Figure 5.1 – National & Organic Industrial Base Roles

While commercial industry drives rapid invention, the DOIB is uniquely positioned to:

- Adapt technologies to real battlefield conditions
- Remove non-essential features and increase ruggedization
- Maintain focus on survivability rather than aesthetics
- Execute modifications in austere environments
- Extend life cycles under extreme conditions
- Preserve continuity during commercial disruption

The DOIB exists inside the military ecosystem. It understands operational reality, Soldier behavior, environmental extremes, and logistical constraints better than any external partner.

That proximity translates into an advantage.

In a GTO (Government-Tested-Only) environment, the DOIB can rapidly:

- Modify commercial systems for combat use
- Strip vulnerabilities
- Harden cyber components
- Localize production when supply chains are threatened
- Continue production if commercial firms fail or withdraw
- Preserve institutional knowledge across generations

The DOIB augments and continues the work of the commercial base.

It is the national insurance policy against:

- Market collapse
- Foreign interference
- Exploited dependence
- Corporate consolidation
- Sanctions or embargoes

In C2G, the DOIB is not a fallback. It is a **parallel pillar of dominance.**

MOSA and the DOIB

MOSA significantly enhances the role of the entire Organic Industrial Base, not just the DOIB. By requiring modular, open interfaces, MOSA allows the DOIB to modernize subsystems, replicate components, and introduce sovereign upgrades without relying on proprietary vendor architectures.

The DOIB can replace modules when commercial supply chains are disrupted, harden components for combat conditions, and continue capability when private firms divest or pivot. MOSA therefore strengthens national resilience by ensuring the DOIB can augment, extend, and continue capability independent of external permission.

5.4 Commercial, Dual-Use and Military-Only Capability Classes

As mentioned earlier, under C2G, capabilities are understood not just by function, but by **origin and intent**. Figure 5.2 demonstrates how they all can contribute to the improvement of combat capability.

Why this matters:
Modern capability does not originate from a single source. C2G integrates commercial, dual-use, and military innovation into one synchronized system.

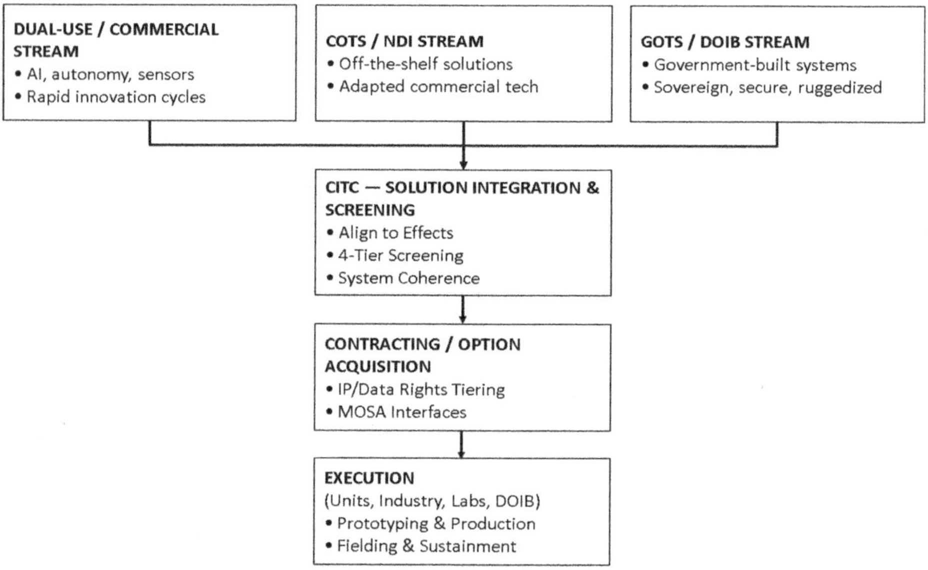

Figure 5.2 – Capability Origins – Commercial, COTS/NDI, GOTS

5.4.1 Commercial and Dual-Use Capabilities

Dual-use technologies serve both civilian and military sectors. Examples historically include:

- GPS
- The internet
- Jet engines

- Composite materials
- Microprocessors
- AI-based decision tools

These technologies are often created in commercial spaces first, then adapted to military use — or first developed for military use and later migrate to civilian markets.

Dual-use is not a risk — it is a **multiplier**.

It accelerates innovation through mass production and civilian competition.

However, it also increases exposure and therefore risk.

C2G mitigates this risk through:

- Enhanced vetting of suppliers
- Cybersecurity validation
- Domestic manufacturing requirements for certain components
- End-to-end supply chain tracking
- Secure microelectronics verification
- Software integrity audits

The goal is not to eliminate commercial engagement.
The goal is to **build trust without surrendering control.**

5.4.2 Military-Specific Capabilities

Some technologies are developed for one purpose only: combat application. These are systems whose exposure and uncontrolled release would carry unacceptable risk.

These capabilities are governed under:

- Tier I intellectual property requirements
- Sovereign production authority

- Restricted supply chains
- Cleared personnel only environments
- Enhanced counterintelligence protection

These systems define strategic advantage and must be protected accordingly.

The C2G model treats classification not as a barrier to innovation — but as a layer of protection around what matters most.

5.5 Supply Chain Vulnerability and Sovereignty Risk

The modern world is interconnected. Components may originate from dozens of countries, pass through untraceable hands, or be produced under unknown conditions.

This creates vulnerabilities including:

- Hidden hardware backdoors
- Embedded malware
- Counterfeit materials
- Disrupted logistics during conflict or crisis
- Economic coercion by foreign states
- Undetected quality degradation

These vulnerabilities turn supply chains into potential attack vectors.

Within the C2G construct, the supply chain is treated as both an **operational and cyber domain.**

Every critical supplier is evaluated through:

- Risk scoring
- Ownership transparency
- Geopolitical exposure
- Cyber hygiene compliance

- Redundancy coverage
- Domestic substitution ability

Suppliers that cannot meet security standards are either:

- Replaced
- Closely monitored
- Or transitioned to the DOIB for replication

This is not isolationism.

It is intelligent resilience.

5.6 Intellectual Property as Industrial Sovereignty

No industrial base can be sovereign if it does not control its core designs. MOSA reinforces sovereign control by reducing dependency on proprietary, closed architectures that obscure risk and limit government authority.

Open interfaces make intellectual property boundaries visible, enabling cleaner negotiation of rights and reducing the danger of hidden vendor lock. When paired with Tier I/II/III data-rights classifications, MOSA ensures the nation can retain technical control, enhance security, and transition critical components to the DOIB whenever required.

Under this selected model:

Tier I systems require Unlimited or Government Purpose Rights without exception.

This guarantees that:

- The nation can continue production
- The DOIB can assume manufacturing
- Allies can be supported if necessary
- Adversaries cannot leverage dependency

Tier II and Tier III systems allow for negotiated rights — but only if those rights support continuity during crisis.

Every capability, regardless of origin, enters the CER with a **data rights classification** that ensures leadership understands ownership, dependency, and risk in real time.

In this way, intellectual property becomes not just a legal issue — but a **command decision variable**.

C2G therefore turns legal data into operational knowledge.

5.7 Embedding Soldiers Within the Industrial Ecosystem

Chapter 4 discussed the importance of the human element in C2G. It is important to identify their role **within the industrial base**. While opportunities exist for Training With Industry, C2G proposes allowing selected Soldiers — from junior enlisted to senior NCO and officer levels — to embed in:

- Commercial labs
- Prototype facilities
- Testing ranges
- Innovation hubs
- DOIB workshops
- Development teams

Their purpose is not to oversee.

It is to **inform and humanize** the design process.

The benefits are twofold:

1. Civilian designers gain an authentic understanding of the Soldier, removing harmful stereotypes and building empathy.
2. Soldiers gain awareness of civilian expectations, culture, and innovation methods — softening reintegration friction later in life.

Junior personnel are critical here. They possess:

- Direct exposure to current conditions
- Fresh perspective unshaped by bureaucracy
- Deep understanding of equipment shortcomings
- Immediate awareness of human-factor failures

C2G does not sideline them.

It **elevates their voice into the design space**.

This blurs the line between "warrior" and "inventor" — in a way that strengthens both.

5.8 Learning from the Industrial Revolution

There was a time when armies moved only at the speed of human and animal muscle.

The Industrial Revolution transformed warfare through:

- Railroads
- Steam power
- Mass manufacturing
- Internal combustion engines
- Mechanized logistics
- Communications infrastructure

These were not optional improvements.

They were existential requirements.

Armies that failed to adopt industrial capabilities lost relevance. Those that embraced them shaped history.

C2G is this generation's Industrial Revolution.

But instead of steam and steel, the drivers are:

- Data
- Speed
- Autonomy
- Artificial intelligence
- Flexible manufacturing
- Decentralized production

The pattern is the same.

Adapt — or become obsolete.

C2G ensures that adaptation is not reactive.

It is **systematic.**

5.9 Redundancy Without Waste

To the uninformed observer, parallel development within industry or within the military looks inefficient. It can be expensive and duplicative. But it is also one of the greatest mechanisms to speed up development – especially in competitive markets.

To the strategist, redundancy in the industrial base is survival.

C2G embraces **intentional redundancy**, maintaining multiple promising lines of development until a clear performance leader emerges.

This prevents:

- Single-point failure
- Technology monopolization
- Predictability
- Strategic vulnerability

Redundancy also aligns perfectly with Middle Tier Acquisition and OTA mechanisms which reward experimentation and limit commitment until capability is proven.

Where legacy procurement sought certainty, C2G seeks **competitive emergence**.

Just as in natural evolution, multiple variations allow the strongest to thrive.

Chapter 5 Core Truth

An industrial base designed for adaptability and sovereignty creates deterrence before conflict begins.

CHAPTER 6 — ACCELERATION THROUGH STRUCTURE AND CONTRACTING

6.1 Speed as a Strategic Advantage

In modern competition, speed is no longer a supporting attribute.
It is a decisive variable.

Adversaries do not need to be better equipped to win — only **faster to adapt.**

Markets respond in hours. Artificial intelligence evolves in days. Cyber threats adapt in minutes. The battlefield no longer tolerates ten-year solution cycles.

Within C2G, speed is recognized not as recklessness, but as **a disciplined form of courage.**

"Who dares wins" is not about impulsive action.
It is about **decisiveness in uncertainty.**

This shifts leadership expectations.

Where legacy systems rewarded careful delay, C2G rewards **risk-managed decisiveness**—elevated as a leadership competency equal to courage.

Speed in this context is structured. It is informed by data. It is measured by Effect. It is bounded by ethics and protected by governance, but it must also be synchronized.

Speed in this context does not imply recklessness. It reflects disciplined decision-making enabled by visibility, feedback, and ethical governance rather than the bypassing of rigor.

Not all elements of a capability adapt at the same tempo. Software and modular subsystems may evolve on commercial timelines, while platforms and strategic systems necessarily move more deliberately. C2G aligns these tempos rather than forcing them into false uniformity.

Acceleration requires synchronization.

6.2 The Function of the CITC in an Accelerated Environment

The Combat Innovation & Transformation Cell (CITC) may be misunderstood as a potential gatekeeper — a body that filters and blocks ideas before they reach leadership. But it is the structure of the CITC that accelerates innovation.

The CITC does **not** exist to deny innovation.
It exists to **translate it into Effects.**

Its core mission is to:

- Understand the operational Effect being sought
- Align proposed solutions to CER-identified needs
- Filter spam, not creativity
- Structure information for fast leadership decision
- Protect decision-makers from overload — not from ideas
- Ensure visibility, not bureaucracy

As the number of proposals grows, a **four-tier triage framework** is introduced:

Tier A — Immediate operational relevance
Directly addresses gaps recorded in the CER

Tier B — Emerging relevance
Promising but not urgent

Tier C — Future or conceptual relevance
Logged for monitoring

Tier D — Non-aligned or unsuitable
Rejected with feedback

This is not suppression. It is **cognitive bandwidth management**.

The CITC exists to **reduce friction between innovators and decision-makers**, not increase it.

This transforms it from perceived bureaucratic obstacle into an **enabler of clarity.**

6.3 Contracting as a Weapon System

Why this matters:
Contract structures determine how fast ideas become operational reality. In C2G, contracting is not administration—it is leverage.

In C2G, contracts do not exist only to acquire objects.

They exist to **acquire options, speed, leverage, and independence.**

The contract itself becomes a strategic instrument — as powerful as any platform.

Traditional contracting is descriptive.
C2G contracting is **architectural.**

It determines:

- Who controls adaptation
- Who controls data
- Who controls production
- Who controls transfer
- Who controls survivability

Recall that contracts establish Data Rights as discussed in Chapter 2. They are negotiable, but cannot put the nation at risk. **The contract is the vehicle to establish sovereignty.**

6.4 Flexible Contract Structures That Enable Speed

The Adaptive Acquisition Framework (Figure 6.1) has provided multiple contracting pathways to help accelerate outcomes.

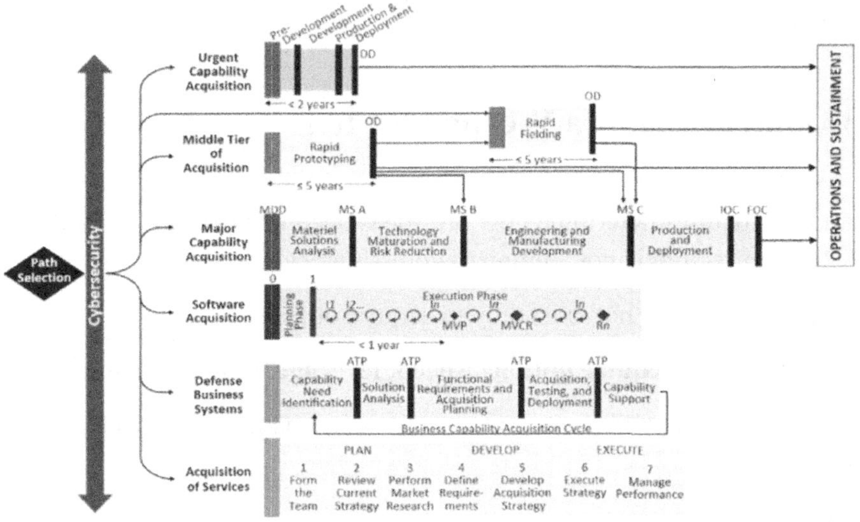

Figure 6.1 – Adaptive Acquisition Framework

C2G builds upon this and leverages a variety of non-traditional and accelerated contract mechanisms, including:

Other Transaction Authority (OTA)

Used for rapid prototyping, iterative testing, and rapid follow-on production when a concept proves viable.

Middle Tier Acquisition (MTA)

Enables fielding of mature capability within 2–5 years, bypassing traditional milestone structure.

Commercial Solutions Offerings (CSO)

Pulls direct innovation from the commercial market with minimal administrative burden.

Broad Agency Announcements (BAA)

Allows concept-first engagement without rigid requirement locking.

Indefinite Delivery Indefinite Quantity & Multiple Award Vehicles

Encourage competition without slowing entry.

Each of these is aligned to the C2G concept of **acceleration and parallel motion** rather than sequential delay.

Under these models, multiple candidate solutions are encouraged — preserving redundancy until a clear winner emerges.

This ultimately serves as a **strategic insurance.**

6.5 Surge and Limited-Adaptation Contract Design

To support rapid scaling and battlefield adaptation, C2G contracts include specific architectural clauses:

- Surge capacity triggers
- Modular scalability provisions
- Geographic manufacturing dispersal
- DOIB transfer readiness
- Rapid licensing expansion
- Embedded change-authority
- Emergency tooling rights
- Component substitution rights

These clauses eliminate negotiation delay during crisis.

Instead of rewriting contracts in wartime, C2G **designs them for wartime in advance.**

This preserves speed when it matters most.

6.6 Controlling Duplication Without Killing Innovation

In legacy systems, leaders feared duplication of effort.

In C2G, leaders fear **single-path failure.**

The difference is profound.

Duplication is waste when unmanaged.

When affordable, contracting redundancy enables flexibility of development.

Through the CER, leaders maintain full awareness of parallel developments. Instead of shutting them down prematurely, promising pathways are sustained until real-world effectiveness separates the viable from the impractical.

This avoids the catastrophic risk of "betting everything" on the wrong version.

History is filled with examples:

- VHS vs Beta
- Competing aircraft designs
- Competing propulsion methods
- Competing missile architectures

The winner is not predictable at inception.

C2G can leverage contracting to protect the **right to discover the correct future.**

6.7 Cybersecurity and Supply Chain as Contractual Requirements

Every accelerated contract includes embedded cyber and supply chain mandates:

- Zero-Trust architecture requirements
- Continuous penetration testing
- Software Bill of Materials (SBOM)
- Secure microelectronics verification
- Ownership transparency standards
- Subtier vendor disclosure

Cybersecurity is not a technical supplement.

It is a **non-negotiable element of capability.**

Contracts that cannot meet cyber standards do not advance — regardless of performance.

This removes emotion from decisions and ensures logic governs selection and progress.

6.8 Modular Open Systems Architecture (MOSA) as a Contractual Accelerator

Figure 6.2 – MOSA Ecosystem

Modular Open Systems Architecture (MOSA) is one of the most powerful technical enablers of C2G. It provides the structural flexibility required for rapid adaptation by ensuring that systems are built from modular, interchangeable components with open, discoverable interfaces. Requiring this contractually allows subsystems to be upgraded, replaced, or inserted independently — dramatically reducing redesign cost, integration friction, modernization delay, but also vendor lock.

Where C2G aligns decisions around *Effects*, MOSA aligns technology around *interfaces*.

MOSA is critical because it concentrates adaptability at the subsystem and software layers, allowing platforms to remain stable while capability evolves continuously.

This creates a synchronized environment where:

- New capabilities can be inserted without redesigning entire platforms
- Multiple vendors can compete on subsystems instead of whole systems
- COTS, dual-use, NDI, and GOTS solutions can be integrated quickly
- The DOIB can augment or continue production without vendor restriction
- Cyber vulnerabilities can be isolated to modules rather than entire architectures
- Redundancy can be preserved while reducing waste
- Portfolio executives (PAEs/CPEs) can orchestrate cross-platform refresh cycles

MOSA transforms major platforms from static investments into continuously adaptable frameworks.
It ensures that the architecture itself enables speed.

In C2G, MOSA is not simply an engineering preference. It becomes a contractual differentiator and development accelerator.
It is a strategic design requirement that:

- Increases resilience
- Enhances sovereignty
- Accelerates innovation
- Reduces long-term cost
- Expands industrial participation
- Enables Effect-driven modernization

MOSA is the **technical backbone** of adaptability — the architectural expression of C2G.

6.9 Patient Speed and Measured Risk Management

Speed does not mean recklessness.

The C2G structure embraces Limited User Assessments (LUA) and Limited User Tests (LUT) to allow **real-world validation of prototypes** in controlled operational contexts.

Appropriate personnel for this role include:

- Highly experienced operators
- Specialized test units
- Embedded innovation Soldiers
- Mission-exposed junior personnel under controlled supervision

These individuals are selected not simply for rank — but for **judgement, adaptability, and situational awareness.**

Risk acceptance is determined in advance and scaled according to:

- Mission importance
- Maturity of technology
- Environmental conditions
- Contingency capability
- Failure consequences

In this way, speed and responsibility stay aligned.

6.10 Alignment to National Power

Part of the C2G strategy is to align with the elements of National Power. C2G strengthens all major elements of national power simultaneously:

- Military: Enhanced speed and adaptation
- Economic: Strengthened domestic manufacturing and innovation economy
- Informational: Increased understanding and narrative dominance
- Diplomatic: Controlled transfer and collaboration with allies

By integrating commercial and organic capability into a unified strategy, C2G transforms military modernization into **national renewal capacity** without undermining sovereignty.

6.11 Contracting as a Cultural Transformation Tool

Beyond its technical role, contracting in C2G reshapes culture.

It rewards:

- Creativity
- Transparency
- Collaboration
- Accountability
- Speed
- Responsibility

Due to the economic nature of the transaction between the commercial sector, there will always be a transactional instrument required. Contracts are leveraged to maintain identity and prevent ethical violations.

In C2G, vendors are not treated as outsiders.

They are treated as **teammates in national service.**

But the boundaries are clear.

- Partnership: always
- Dependency: never

This clarity removes mistrust and creates a healthier innovation ecosystem for all participants.

Chapter 6 Core Truth

Speed becomes sustainable only when structure, contracting, and risk are deliberately aligned to enable adaptation.

CHAPTER 7 — DOTmLPF-P IN A C2G WORLD

Doctrine, Organization, Training, materiel, Leadership, Personnel, Facilities, and Policy (DOTmLPF-P) are not checkboxes. In C2G, they are levers of speed — or anchors of delay.

Figure 7.1 demonstrates how the tenants of DOTLmPF-P both drive and are driven from the military and commercial sectors.

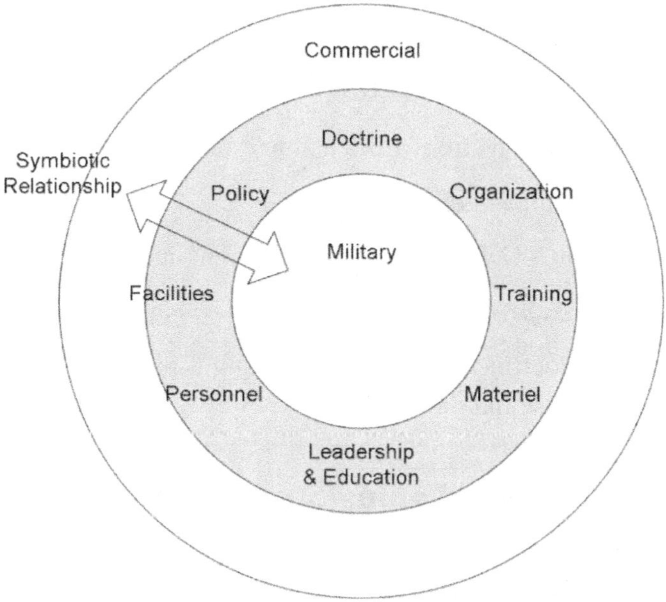

Figure 7.1 – DOTmLPF-P Connects Military to Commercial

7.1 Doctrine: From Static Guidance to Living Framework

Why this matters:
Transformation often fails inside DOTmLPF-P. C2G turns it from a compliance checklist into an engine for coordinated change.

Doctrine has long provided stability, clarity, and shared understanding across generations of military professionals. It establishes common language, expectations, and alignment.

However, doctrine built only for permanence can lag a world defined by change.

In a C2G environment, doctrine does not disappear — it evolves.

Rather than serving as a fixed prescription, doctrine becomes a living framework, continuously informed by:

- Operational experience
- Emerging technologies
- CER-captured Effects
- Lessons from allies and partners
- Human-centered feedback

Doctrine in a C2G world answers a different question than before:

"What must be preserved, and what must change?"

In this way, doctrine becomes a stabilizing force in an accelerating world — not a constraint on it.

7.2 Organization: Aligning Structure to Speed

Traditional military organizations are optimized for clarity of responsibility and unity of command. These principles remain essential.

However, C2G introduces the need for structures that are both disciplined and adaptive.

This requires:

- Cross-functional teams aligned around Effects, not branches
- Fluidity between operational and developmental roles
- Formal connection points to the CER and CITC
- Space for innovation without fragmenting authority

Organizational design in a C2G environment favors:

- Mission-centric alignment
- Rapid information flow
- Reduced lateral friction
- Transparency of purpose
- Clear, bounded authorities

Execution occurs in the limbs — units, labs, industry, and the DOIB — (further discussed in Chapter 9) but the nervous system must be redesigned for speed.

Centralized coordination with decentralized execution produces adaptive power.

This reflects historical lessons such as ***auftragstaktik*** (mission-type orders) while remaining firmly grounded in modern mission command philosophy. It takes operational employment methodologies and applies it to acquisition and capability generation.

7.3 Training: Embedding the Future into the Present

If capability must be generated faster, then learning must be embedded earlier.

Training within the C2G construct is no longer confined to:

- Classrooms
- Ranges
- Simulators
- Formal schools

Training becomes a living, embedded experience within the full capability ecosystem:

- Soldiers embedded within innovation cells
- Technologists embedded within units
- Designers embedded within operational environments
- Decision-makers exposed to consequence-based feedback loops

This creates a force that learns not only how to use new capability — but how to think about generating it.

Training within C2G includes:

- Systems thinking
- Ethical application
- Effects-based reasoning
- Risk awareness
- Supply chain consciousness
- Cyber temperature checks
- Decision discipline
- Adaptation literacy

In this way, training is no longer just preparation for war.

It is preparation for change.

The training domain becomes the engine room of adaptation.

7.3.1 The Second KISS Principle: Keep It Similar, Stupid

Many developers have heard of the 'KISS' principle – Keep It Simple, Stupid. This principle assures rapid acceptance and intuitive interfaces for the user.

However, as new capabilities are introduced, C2G embraces a corollary to the original KISS maxim: **Keep It Similar, Stupid**. When designing or adopting new technology, maintaining continuity with familiar interfaces, workflows, and cognitive patterns dramatically reduces training burden and accelerates adoption. When Effects can be achieved through technologies that build on existing user experience, the force transitions faster and with greater confidence.

Innovation should not be alien when familiarity achieves the same effect.

Example: Qualifying the Apache as an Instrument Flight Rules Aircraft

When the AH-64 was developed, it was designed to detect and engage enemy forces including at night. Weather phenomenon such as fog or clouds limited operations. Instrument flight was not initially required. While the Apache included basic instruments for recovery from Inadvertent Instrument Meteorological Conditions (IIMC), it did not support routine instrument flight operations.

When instrument flight capability was later introduced, two approaches emerged. Some experienced pilots favored evolving existing Apache symbology. Others – including junior aviators –recommended adopting a Federal Aviation Administration (FAA) standard flight display.

To reduce fleetwide risk, both solutions were fielded.

This allowed pilots to train from the beginning on FAA-standard symbology, while experienced aviators adapted through incremental change – achieving the same operational effect with minimal disruption and reduced risk.

While this case demonstrates the value of disciplined, incremental adaptation within a stable governance framework, it does not guarantee that similar outcomes can be replicated indiscriminately, effectively, or efficiently.

7.4 Materiel: From Assets to Effects Enablers

In a C2G framework, materiel is not sacred. It is not a symbol. It is not an identity marker. It is not a legacy.

It is a tool — valuable only in proportion to the effect it enables.

This perspective allows for:

- Shorter lifecycles
- Rapid modification
- Hybrid solutions
- Modular design
- COTS, NDIs, and GOTS in an integrated ecosystem

The DOIB is not in competition with the commercial base.

It augments and continues capability to ensure:

- Sovereignty
- Continuity
- Resilience
- Security
- Independence of decision

Materiel becomes a means, not a master.

MOSA as a Materiel Enabler

MOSA reinforces the C2G principle that materiel is valuable only insofar as it enables effects. A modular, open architecture ensures materiel can evolve at the pace of operational demand. Instead of replacing entire platforms, the military upgrades components, refreshing capability without discarding the underlying system. This prevents materiel from becoming a constraint and allows Effects to remain synchronized with the CER as missions evolve.

7.5 Leadership: Courage Reframed as Intelligent Decisiveness

In traditional language, courage is often associated with physical risk.

In a C2G environment, courage also includes risk-managed decisiveness.

This form of courage includes:

- Choosing speed over comfort
- Accepting innovation over familiarity
- Embracing uncertainty with discipline
- Making decisions in incomplete information environments
- Standing behind logical judgments
- Defending necessary change

This is not reckless ambition.

It is informed boldness — and it deserves recognition as a formal leadership competency equivalent to physical courage.

In C2G, hesitation is not neutral.
It creates vulnerability.

Leadership must therefore be taught, rewarded, and evaluated based on the ability to decide clearly, adapt quickly, and hold ethical ground.

7.6 Personnel: Blended Experience and Permeable Boundaries

C2G acknowledges the reality that the best ideas do not always respect uniform boundaries.

C2G does not erase those boundaries.

It allows permeability without dilution.

Personnel structures in a C2G system should allow:

- Temporary cross-assignments
- Dual-exposure rotation models
- Embedded civilian-military exchange
- Industry immersion opportunities
- Return-to-force pipelines

This does not blur the line between civilian and Soldier.

It acknowledges that:

- The obligations of military service are unique
- The authorities of military command are distinct
- The responsibilities borne by Soldiers are unequaled

Selective crossover strengthens both worlds when handled deliberately.

C2G therefore treats people as assets of perspective – not just position.

7.7 Facilities: From Static Sites to Adaptive Nodes

Facilities within a C2G model shift from monolithic structures to adaptive nodes within a network.

These nodes may include:

- Innovation outposts
- Test environments

- Co-location spaces
- Cyber-secure collaboration hubs
- Mobile laboratories
- Deployed development environments

Rather than centralizing everything geographically, C2G encourages distributed intelligence with unified direction.

Facilities become places where:

- Ideas meet reality
- Data meets discernment
- Humans meet systems
- Vision meets action

They are not just buildings.

They are capability accelerators.

7.8 Policy: The Quiet Enabler of Speed

Policy is often seen as restriction.

In C2G, it becomes permission.

The goal is not the absence of policy, but the right policy — intentionally written to:

- Encourage innovation
- Protect sovereignty
- Enable experimentation
- Control risk without suffocating progress
- Support rapid contracting
- Clarify authority

Well-written policy does not slow progress.

It directs it.

This requires close connection between:

- Strategic leadership
- Legal advisors
- Technologists
- Operators
- Contracting Officers
- Industrial partners

Policy becomes the architecture of safe speed. Not a brake. A steering wheel.

Chapter 7 Core Truth

DOTmLPF-P elements either accelerate adaptation or constrain it, depending on how deliberately they are aligned.

CHAPTER 8 — PILOT, SCALE, AND COMPETITIVE OVERMATCH

8.1 Speed with Structure - Tempo

Speed is frequently misunderstood. The phrase "Slow is smooth and smooth is fast" captures the paradox, illustrating that tempo – the rate at which activity is executed – accelerates outcomes when properly managed.

In C2G, speed is *disciplined.*

It is structured.
It is measured.
It is intentional.

It remains urgent.

True speed is not determined by how rapidly action is taken, but by how quickly relevance is established and advantage is secured.

To move fast without structure is to generate chaos.

To move with structure – guided by purpose, logic, and ethical grounding – is to generate power.

C2G does not replace deliberation.
It compresses unnecessary delay and removes friction — not reasoning.

Speed, therefore, is not the enemy of strategy.

It is the result of clarity.

8.2 The Role of the Pilot Phase

The pilot phase is the transition point between theory and reality.

It is where:

- Concepts meet constraints
- Design meets environment
- Intention meets consequence
- Hypothesis becomes evidence

In traditional constructs, pilot programs are either rushed and ineffective or they seek perfect conclusions and can take years – often losing their relevance before results are produced.

Under C2G, pilots are deliberately designed to be:

- Limited in scope
- Rapid in execution
- Focused on Effect
- Policy-protected
- Resource-disciplined
- Measurement-driven

The purpose of a C2G pilot is not perfection.

It is risk-managed learning.

It asks:

- Does this concept move reality in the intended direction?
- Does it create the desired Effect, even if imperfectly?
- Does it present unacceptable risk?
- Does it reveal constraints not previously understood?

A pilot is not a verdict.

It is a conversation with reality.

8.3 Learn Fast: The Discipline of Accelerated Insight

Earlier innovation models often praised the idea of "failing fast."

C2G evolves this language.

Failure may produce learning, but learning does not require failure.

The discipline is not in celebrating breakdown.

It is in accelerating understanding.

"Learn fast" means:

- Observing without ego
- Documenting without distortion
- Adjusting without attachment
- Improving without delay
- Preserving what works
- Releasing what does not

This approach requires intellectual humility — the willingness to accept that even elegant designs may not suit reality.

In C2G, learning is not informal.

It is structured into:

- Data collection
- Embedded observation
- User feedback loops
- Technical analysis
- Ethical evaluation
- Environmental influence

Learning is treated as an **operational discipline**, not a side effect.

8.4 Measuring Effect Without Reducing Humanity to Metrics

Why this matters:
What is measured determines what is valued. C2G measures speed, risk, and resilience without reducing people to expendable variables.

C2G requires measurement. Without it, effectiveness becomes guesswork.

But measurement in C2G is not about treating people as numbers.

It is about understanding changes in:

- Risk exposure
- Time-to-impact
- Decision efficiency
- Operational burden
- Sustainability
- Resilience
- Deterrence credibility

These are measured without trivializing the human condition — without forgetting that behind every metric stands a human being who bears its weight.

The purpose is to determine whether stability and effectiveness are increasing across the C2G enterprise.

Measurement is not performed in isolation or selectively.

It is conducted holistically to confirm accelerated adaptation and development across the system.

8.5 Competitive Overmatch: Out-Thinking and Out-Adapting

During the Cold War, dominance was often achieved through mass and economic endurance. The United States out-spent and out-produced adversaries whose systems could not keep pace.

That era has shifted.

Today, advantage is not determined only by who produces more – but also by who adapts faster.

Competitive overmatch in the C2G framework means:

- Anticipating change before others recognize it
- Seeing patterns before they form
- Aligning resources before urgency appears
- Adapting faster than competitors can innovate
- Decentralizing creativity while centralizing purpose

This is not about escalation.

It is about rendering opposition irrelevant.

This is not dominance through destruction.

It is dominance through adaptability.

8.6 Scaling Without Distortion

Scaling is dangerous.

Systems that work beautifully when small often collapse when expanded.

Pressure, politics, personalities, and legacy structures distort pure design.

C2G approaches scaling as a progressive discipline – not uncontrolled expansion.

Scaling must preserve:

- Integrity of intent
- Ethical alignment
- Human-centered design
- Speed of learning
- Quality of data
- Coherent governance

Instead of "replicating the pilot everywhere," C2G asks:

- Where does this matter most right now?
- Where does it add the greatest value?
- Where does it reduce the greatest risk?

This phased expansion allows the system to breathe as it grows – rather than fracture under its own size.

8.7 Centralized Insight, Decentralized Execution

All effective military systems recognize decentralized execution under centralized intent. This applies also when piloting and scaling adaptations.

C2G extends this principle beyond operations and into innovation.

The continued centralized focus, but decentralized execution of pilot programs and scaling applies ***auftragstaktik*** once again outside the operational domain. Empowering subordinates and functional commands becomes critical to success.

As a reminder, within C2G:

- The CER defines intent
- The CITC aligns direction
- The PAE/CPE ensures coherence
- The limbs (units, labs, industry, DOIB) execute according to reality

This structure allows:

- Flexibility at the edge
- Stability at the center
- Speed without fragmentation
- Accountability without rigidity

Execution becomes an act of distributed intelligence.

8.8 Competitive Advantage as a Deterrent

C2G does not seek conflict.

Its ultimate success is the conflicts that never occur.

When a nation demonstrates that it can:

- See change early
- Adapt rapidly
- Generate capability responsibly
- Protect its people
- Maintain legitimacy
- Preserve humanity

…then the desire to challenge it diminishes.

Deterrence is not the announced outcome but it is an achieved byproduct from visible competence and disciplined application.

Chapter 8 Core Truth

Learning faster than an adversary scales advantage more reliably than perfect execution.

CHAPTER 9 — GOVERNANCE, ALIGNMENT, AND THE CAPABILITIES & EFFECTS REGISTRY

9.1 Why Governance Must Change for Speed to Exist

Legacy acquisition governance was built around programs, budgets, and platforms. It rewarded control through oversight layers, approvals, and reviews. That model was effective when change was slow and threats were predictable.

C2G requires a different form of governance.

Not less control — but **smarter control**.
Not fewer leaders — but **clearer roles**.
Not more rigidity — but **bounded freedom**.

Governance, under C2G, must be designed to answer three critical questions:

1. Who owns the Effect?
2. Who is responsible for alignment?
3. Who enables execution?

Traditional governance often answered only the third.

C2G demands an architecture that governs the *intent*, not just the *implementation.* Recent acquisition reforms help align all three.

This is how speed and discipline coexist.

9.2 The Capabilities & Effects Registry (CER) as the Central Nervous System

Why this matters:
This is where governance stops slowing speed and starts protecting it. The CER makes rapid adaptation coherent, ethical, and sustainable.

The **Capabilities & Effects Registry (CER)** is one of the most important elements of the C2G construct. As mentioned, it is a **living system of understanding.** But more importantly, it becomes a method of governance and alignment.

The CER becomes the single repository for:

- Recorded operational Effects
- Validated capability gaps
- Approved problem statements
- Mapped solutions
- Redundancy awareness
- Interdependencies
- Technology maturity
- Intellectual property classification
- Supply chain exposure
- Cyber vulnerability posture
- DOIB readiness

Every submission entering the system — from Soldier, unit, partner, lab, or vendor — is viewed **through the lens of Effect**, not through organizational bias. It is also controlled within the CER.

The CER does not exist to judge ideas.
It exists to understand what the battlefield requires and to **translate lived experience into strategic clarity.**

As such, it becomes the anchor point for:

- Program Acquisition Executives (PAEs)
- Capability Portfolio Executives (CPEs)
- CITC operations
- Senior leader understanding
- Long-term strategy
- Near-term adaptation

In C2G, **nothing is hidden in spreadsheets or stovepipes.** If it matters, it exists in the CER.

9.3 Who Enters Effects into the System — And Why Authority Is Distributed

With the volume of potential Effects or Capabilities that could be entered into the CER, a natural concern is whether the CER centralizes too much power at high strategic levels such as the J8 or top resourcing elements?

As discussed earlier, C2G avoids that trap through **distributed input and shared ownership.**

Effects are entered into the CER by:

- Operational commanders
- Junior Soldiers and NCOs
- Embedded personnel in labs and industry
- Analysts within the CITC
- CPE-aligned capability analysts
- Test and evaluation commands
- Coalition partners (when appropriate)

This broad contributor base ensures that **no single organization controls the narrative.**

The authority to validate and align an Effect does not equal the authority to fund or execute it.

This separation prevents single-point suppression or distortion.

A key example:
If an anti-tank capability is identified, it is assigned to the **Effect of enemy armored neutralization**, not to the platform domain of "aviation" or "ground maneuver."

This prevents one portfolio decision from unintentionally destroying an entire category of innovation.

Remember, in C2G, Effects are owned by **the mission — not by a branch.**

That distinction eliminates bias and protects the force from internal competition erasing external advantage.

9.4 The Governance Role of the CITC

The **Combat Innovation & Transformation Cell (CITC)** may appear as an additional level of bureaucracy in terms of governance.

In C2G, it is the opposite.

The CITC is positioned to **enable vision and expedite vision, not deny it.**

Its mandate is to:

- Translate incoming ideas into "Effect language"
- Align them against CER gaps
- Structure complexity for senior leaders
- Protect leaders from overload — not from ideas
- Provide feedback loops to contributors
- Preserve transparency

As proposal volume increases, the CITC applies a structured four-tier triage model (Figure 9.1) to preserve speed, transparency, and decision quality.

Tier A submissions align directly to urgent, CER-recognized Effects and move immediately toward execution pathways. Tier B concepts show emerging relevance and are monitored as conditions mature. Tier C submissions are logged for observation and future reassessment. Tier D submissions are rejected with explicit feedback, ensuring ideas are closed deliberately rather than silently discarded.

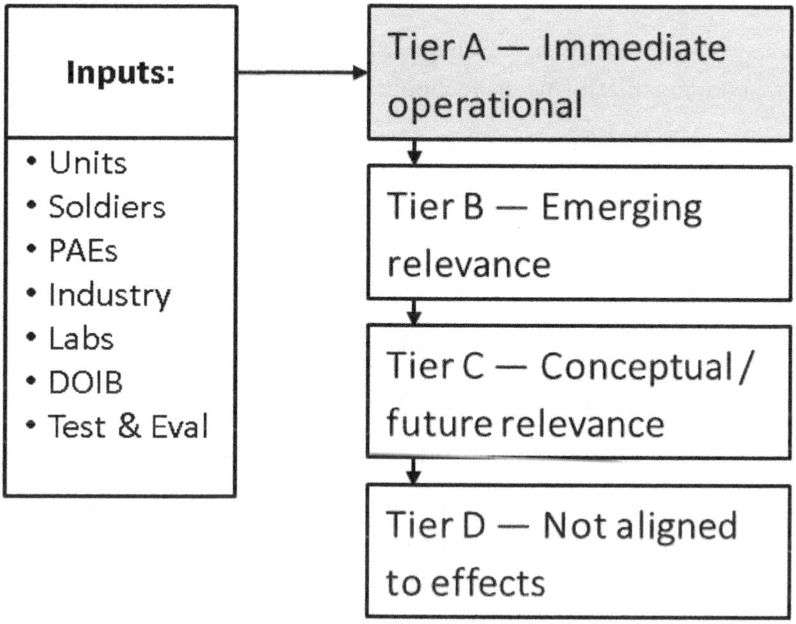

Figure 9.1 – Four-Tier CITC Intake and Screening

Importantly, the CITC's mission is to understand Effects—not to dismiss ideas. This triage structure is intended to preserve speed, transparency, and decision quality at scale, not to suppress ideas or centralize innovation authority. In doing so, it creates a collaborative environment in which innovators feel heard and leaders trust what they see.

9.5 Alignment with Acquisition Transformation, PAEs, and CPEs

The recent Acquisition Transformation guidance and shift toward **Capability Portfolio Executives (CPEs)** aligns directly with the C2G model.

C2G does not compete with this movement.

It **reinforces and operationalizes it.**

Under C2G:

- PAEs become shepherds of Effects, not just overseers of budgets, platforms, or technologies
- CPEs execute acquisition to deliver Effects
- The CER provides the shared awareness platform across portfolios
- The CITC enables cross-portfolio coordination
- Redundancy is visible and intentional
- Synergy is intentional

This layered architecture enables:

Centralized planning with decentralized execution – modern Mission Command and empowerment at the point of action with clarity of intent.

C2G is not an invention of a new philosophy.

It is the digital and institutional expression of a timeless truth:

Clear intent + empowered execution = speed of success

Only now, that success is measured in Effects, not paper milestones.

9.6 Execution Occurs in the Limbs, Not the Brain

It is essential to understand: The CER and CITC do **not** execute – they coordinate.

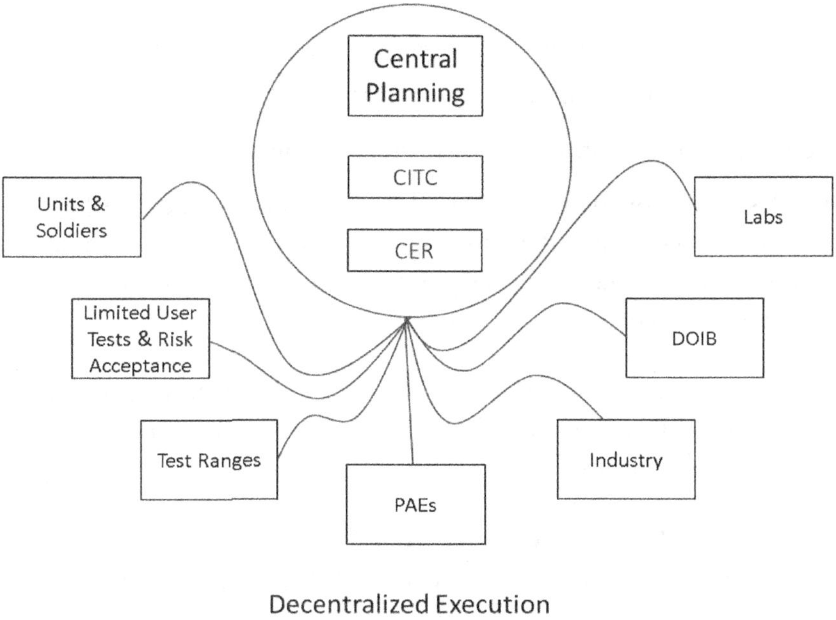

Figure 9.2 – Execution in the Limbs

Execution occurs in the **limbs** of the organization:

- Units and formations
- Laboratories
- Industry partners
- Test ranges
- The DOIB
- Maintenance and sustainment facilities
- Forward environments

These are the hands of the C2G system.

They are decentralized by design to ensure:

- Resilience against disruption
- Local adaptation and understanding
- Mission-focused innovation
- Reduced vulnerability to centralized failure

The headquarters coordinates while the limbs **act.**

This design mirrors biology and successful militaries alike. Just as organizations with distributed control can sense, adjust, and act without waiting for central instruction, C2G's "limbs" operate with localized autonomy while remaining aligned to a single unifying intent.

In the same way, C2G empowers units, labs, industry teams, and the DOIB to act immediately at the point of friction, while the CER and CITC maintain the central purpose.

This **balance of centralized intent and decentralized execution** is what gives the overall system its speed, adaptability, and survivability.

9.7 Intellectual Property, Data Rights and the CER

The CER also maintains visibility as a **sovereignty dashboard.**

Every proposed solution is displayed with:

- Data rights classification (Tier I / II / III)
- IP ownership structure
- Government control level
- DOIB transition readiness
- Allied transfer permissions
- Reproduction authority
- Cyber exposure surface

This allows leaders to see not only what a capability can do — but **who truly controls it.**

If a solution restricts national independence, that risk is visible in real time.

This makes intellectual property a strategic decision variable at the highest levels — exactly where it belongs.

9.8 Ethics, Law, and Legitimacy Built into Governance

Speed without ethics becomes danger.

Therefore, the CER governance model also integrates:

- Law of Armed Conflict (LOAC) considerations
- Rules of Engagement implications
- Civilian harm risk assessments
- Misuse potential
- Information warfare consequences
- Narrative implications

C2G governance is not about winning at any cost.

It is about preserving legitimacy while ensuring superiority.

This is essential for:

- Domestic trust
- Allied confidence
- Post-conflict stability
- Moral clarity

And ultimately — for **deterrence**.

A visibly ethical capability system discourages conflict rather than inviting it.

9.9 The CER as a Deterrence Mechanism for Humanity

While the purpose of this book is not to encourage conflict, one reality cannot be ignored:

An effective C2G framework reduces the likelihood of conflict through **credible deterrence.**

It removes ambiguity.
It removes temptation.
It removes perceived weakness.

By demonstrating that the nation can adapt faster than any rival, war becomes less attractive.

In that sense, C2G does not exist only for Soldiers.

It exists **for humanity.**

A world where aggression is futile is a world where lives are saved — before conflict begins.

Chapter 9 Core Truth

Effective governance enables speed by aligning authority, visibility, and accountability to mission-owned Effects.

CHAPTER 10 — RISK, RESILIENCE, AND RESPONSIBILITY

Specialization Without Adaptability: A Dinosaur Lesson in Strategic Risk

Many dinosaur species were perfectly tuned to specific, narrow ecological niches. This evolutionary specialization gave them tremendous short-term advantage. This can be viewed the same way when defense programs are optimized for a single threat scenario and often outperform broader, multifunctional designs.

But specialization comes with inherent fragility. When the environment abruptly changed, the very designs that once ensured dominance became liabilities.

Modern military capability faces the same risk. Systems built for one environment may falter in another. Doctrines that solve yesterday's problems can harden into vulnerability. Acquisition pipelines optimized around a predictable threat may collapse under chaotic conditions. C2G confronts this reality directly by ensuring that specialization never becomes strategic brittleness.

The lesson is simple:
Success in one era is not insurance for the next. Adaptation is the only safeguard against irrelevance.

10.1 The Cost of Delay in a Networked World

Why this matters:
In a networked world, delay is no longer neutral. Delay creates vulnerability, invites competition, and erodes deterrence.

In earlier eras, a delay in fielding a capability might result in reduced efficiency or a missed opportunity. In the modern world, delay produces cascading consequences.

Every month of inertia risks:

- Strategic surprise
- Loss of technological relevance
- Erosion of deterrence
- Increased vulnerability of allies
- Economic disadvantage
- Loss of global influence

The strategic cost of delay is not always visible, but it is always real. A vacuum of capability invites competitor development, destabilizes power balances, and encourages opportunistic action.

C2G exists to reduce this drag. It is not built to rush blindly, but to **remove unnecessary friction and indecision** from systems that can no longer afford it.

Speed is not for convenience. Speed is for stability.

10.2 Learning as an Accelerant

C2G does not endorse reckless failure.
It endorses **intentional, accelerated learning.**

The operating principle is simple:

Learn fast — and act on what is learned.

Rapid feedback loops, limited user assessments, experimental pilot programs, and iterative refinement are structured to extract insight as quickly and safely as possible. Each test, each prototype, and each early deployment is designed not just to prove success, but to increase understanding.

In C2G, learning is not a byproduct.
It is the primary asset.

Through disciplined analysis of both success and shortcomings, the system evolves faster than any adversary can anticipate.

This creates an asymmetric advantage without requiring escalation.

Adaptation becomes the deterrent.

10.3 Intellectual Property as a Security Vulnerability

In the digital age, intellectual property is no longer just a business asset.

It is a **security surface.**

Stolen designs, compromised source code, or replicated manufacturing techniques can degrade or eliminate a nation's advantage faster than any kinetic attack.

In C2G, intellectual property risk is treated as a form of cyber risk.

Each system is evaluated for:

- Unauthorized access risk
- Replication vulnerability
- Dependency on foreign-held design authority
- Unknown code exposure
- Obsolescence vulnerability

This is why Tier I systems require **Unlimited or Government Purpose Rights without exception**. It is not about corporate distrust; it is about national continuity.

IP control becomes an extension of sovereignty.

And sovereignty is not negotiable.

10.4 Cybersecurity as an Integrated Design Requirement

Cyber threats are no longer limited to networks.

They extend into:

- Manufacturing equipment
- Firmware
- Digital twins
- Logistics systems
- Quality control software
- Supply chain management platforms
- Embedded chipsets

A system is only as secure as its most vulnerable subcontractor.

Therefore, C2G embeds cyber requirements at every level:

- Zero-trust architecture principles
- Continuous threat monitoring
- Software Bills of Materials (SBOMs)
- Third-party certification
- Secure firmware validation
- Tamper-resistant hardware design
- Code-review audit requirements

Cybersecurity is not a box to check.

It is an **operational condition of acceptance.**

If a system cannot defend itself in the cyber domain, it does not move forward — regardless of performance.

10.5 Cultural Friction: The Hidden Adversary

Not all resistance to change comes from outside the organization.

Cultural friction is one of the most powerful and underestimated adversaries.

This manifests as:

- "We've always done it this way"
- Budget protection behavior
- Platform loyalty
- Rank-based inertia
- Excessive process adherence
- Fear of disruptive change
- Bias against non-traditional partners

C2G must confront this friction respectfully, but firmly.

It does not erase tradition. It evolves it.

By embedding Soldiers in innovation, using cross-functional teams, connecting junior operators with senior developers, and creating transparency through the CER, C2G steadily weakens institutional resistance without creating internal conflict.

Culture is not changed by policy alone.

It is changed by **shared experience and proven results.**

10.6 Ethics, Law, and Moral Legitimacy

Speed without moral grounding is dangerous.

C2G integrates ethical considerations throughout the generation process, including:

- Compliance with the Law of Armed Conflict
- Alignment with Rules of Engagement
- Protection of civilian populations
- Accountability for autonomous or semi-autonomous systems
- Transparency in testing, deployment, and intent
- Safeguards against misuse or abuse

This is not a superficial addition.
It is foundational.

A capability that violates ethical standards, no matter how advanced, undermines legitimacy and threatens long-term stability.

The ultimate purpose of C2G is not destruction.

It is **deterrence, protection, and preservation of life through operational overmatch.**

A capability that prevents conflict is more valuable than one that wins it.

10.7 Scalability, Collaboration, and Allied Integration

Even though scalability can be dangerous, C2G is designed to be inherently scalable.

Because the system is Effects-driven, modular, and transparent, it can be expanded:

- Across services
- Across agencies
- Across allied nations
- Across coalition operations
- Across industries
- Across future domains

Allied participation is enabled through:

- Controlled data rights transfer
- Tiered access models
- Shared standards for Effects classification
- Secure collaboration environments
- Interoperability-first design principles

This allows the model to grow without compromising sovereignty.

C2G therefore becomes a shared framework for resilience — trusted, adaptable, and collectively strengthened.

10.8 Risk as a Managed Variable, Not an Avoided One

Risk is not eliminated under C2G.

It is **confronted consciously.**

Every pilot program, every limited deployment, every new integration includes a predefined risk assessment considering:

- Technological maturity
- Operational consequence
- Recovery capability
- Ethical implication
- Cyber exposure
- Strategic impact

This is where leadership judgment becomes essential. It requires the decision of whether to accept acquisition risk or operational risk.

Not all risk is equal.
Not all risk is acceptable.

C2G teaches leaders to ask not, "Is there risk?"

But rather:

"Is this the right risk, at the right moment, for the right Effect?"

That is the difference between recklessness and strategic courage.

10.9 Reducing Risk Through Distributed Capability

C2G is designed for resilience. It comes from a system of systems that provides coverage and reinforcement as needed.

It comes from:

- Redundancy
- Modularity
- Geographic dispersion
- Multiple vendors
- Multiple innovation pipelines
- Distributed authority
- Embedded human adaptability

No single point of failure should ever exist.
No single organization should ever create complete vulnerability.

This structure mirrors the natural world, the digital world, and the most successful military doctrines in history.

It ensures that disruption never creates defeat and ensures manageable risk.

10.10 The Strategic Purpose Beyond C2G

While C2G strengthens readiness, it ultimately serves a higher purpose. It was developed as a framework to enable responsible stewardship of limited resources and safeguard humanity.

C2G helps adapt capabilities faster than threats so:

- War becomes less attractive
- Instability becomes temporary
- Aggression becomes irrational
- Cooperation becomes more likely

In this way, C2G is a responsible "war system" but also a **conflict-prevention engine.**

By demonstrating overwhelming adaptive capacity, C2G encourages negotiation over violence.

It doesn't threaten life. It protects it — through deterrence.

Chapter 10 Core Truth

Risk cannot be eliminated in adaptation, but it can be structured to increase resilience rather than vulnerability.

CHAPTER 11 — THE FUTURE OF C2G

11.1 C2G as a Discipline, Not a Reaction

Why this matters:
C2G is not a program or reform initiative. It is a discipline for surviving and competing in an environment where change never stops.

C2G is not a reaction to a single event, adversary, or era. It is a disciplined framework — a deliberate and enduring approach to transforming intent into Effect, and risk into readiness.

Where legacy systems were structured around conflict response, C2G is designed around **continuous capability creation**.

It is not built solely for times of war.
It is built for **perpetual relevance**.

By unifying requirements generation, technological development, industrial mobilization, human integration, governance reform, and ethical discipline, C2G replaces episodic modernization with **persistent evolution**.

It is a way of thinking. It creates a culture of progress.

And a positive culture changes outcomes for the better.

11.2 Rendering Conflict Unnecessary or Decisive If Pursued

C2G is designed to optimize the generation of combat capability. That is the crux of the model and the focus of this work.

However, the unintended — and profoundly valuable — underlying product of such capability is **deterrence**.

In this way, C2G operates with two outcomes, both aligned to stability:

1. **Conflict becomes unnecessary** because advantage is unmistakable

2. **Conflict becomes decisive** if it is ultimately pursued

As stated previously, deterrence is not the goal of C2G, but it is the natural consequence of dominance through adaptability.

The goal remains unwavering:

To generate capability faster than any adversary can respond.

11.3 The True Strategic Advantage

In a world of uncertainty, no nation can permanently own a specific technology. Knowledge diffuses. Competitors emerge. Innovation spreads.

What cannot be copied as easily is a **system that learns, adapts, integrates, and regenerates itself continuously.**

Resilience is not a trait of hardware. It is a property created by a holistic system and culture that spans:

- Governance
- Leadership
- Training
- Industrial design
- Contracting
- Ethics
- Collaboration
- Feedback loops
- Distributed execution

This means that even if one generation of technology becomes obsolete, the **ability to generate the next is never lost.**

That is where true strategic advantage resides.

Not in possession — but in regeneration.

11.4 The Universal Pattern of Adaptation

Although born in a military context, the pattern that underlies C2G is not exclusive.

It is, in fact, seen in the natural world as **adaptation and evolution.** The battlefield simply forces clarity. But the principles are universal.

Systems that survive do not do so because they are perfect.
They survive because they adapt faster than the conditions around them change.

The same is true of:

- Organizations
- Economies
- Ecosystems
- Technologies
- Cultures
- Species
- Civilizations

C2G is nothing more — and nothing less — than **a tailored model of adaptation applied with intention.**

This makes it **appropriate for tailoring to create other effective organizations.**

11.5 Investing in Humans

C2G intentionally embeds human experience into every layer — not just as users, but as contributors and co-creators.

When Soldiers are embedded in development environments, and civilians are brought closer to service members, a powerful transformation occurs:

- Society better understands its protectors
- Soldiers better understand the world they return to
- Innovation becomes empathetic
- Ethics become intimate
- Transition becomes less traumatic
- Purpose becomes shared

The result is a system that serves humanity, not just a military function.

This is an extraordinary by-product.

But it is not accidental.

It is embedded by design.

11.6 The Strategic Value of Learning Over Perfection

C2G is not a pursuit of flawless execution.

It is a pursuit of **relentless improvement.**

"Learn fast" becomes the enduring principle — not as permission to be careless, but as discipline to extract wisdom quickly and apply it without ego.

In C2G, value is found not in avoiding error, but in **converting experience into knowledge at speed.**

It becomes a renewable resource, immensely powerful, and distinctly human.

11.7 A Model That Extends Beyond the Military

Although this book is grounded in combat capability, its architecture extends naturally into other domains that require rapid, ethical, resilient adaptation.

This invites the concept of **Universal Capability Generation**:

A model that can be applied to:

- Disaster response
- Climate adaptation
- Public safety
- Medical advancement
- Infrastructure modernization
- Space and exploration
- Energy independence
- Global resilience

They represent **the next horizon.**

And they provide the foundation for a future work that moves beyond the military while preserving everything this one establishes.

11.8 The Final Truth

Weapons do not win wars.

Systems do not save nations.

People do.

And the greatest protection we can offer people is not superior firepower — but a framework that keeps conflict from occurring in the first place.

C2G, when properly executed, transforms fear into stability, uncertainty into resilience, and chaos into thoughtful preparedness.

It does so quietly.
It does so decisively.
It does so ethically.

And most of the world will never realize it is happening.

That may be its greatest success.

Chapter 11 Core Truth

Organizations that institutionalize learning outperform those that pursue perfection.

Chapter 12 Conclusion — The Courage to Dare

C2G is not a proposal for abstract reform. It is the start of a framework that recognizes the environment as it exists and changes.

The character of competition has changed. Threats evolve faster. Technology proliferates faster. Adversaries observe, adapt, and exploit at a pace that renders slow, platform-centric capability generation increasingly fragile. At the same time, the United States cannot pause operations in order to modernize. The force must remain ready, credible, and engaged while continuously adapting. The tension between urgency and responsibility is the defining challenge of modern military capability.

C2G exists to resolve that tension.

At its core, C2G is a framework that enables a unified, Effects-driven system that connects operational need to delivered capability through integrated governance, decentralized execution, and disciplined adaptation. It replaces fragmented processes with a logical coherent enterprise. It shifts focus away from platforms as ends unto themselves and toward the Effects that matter in real operations. It concentrates decision authority where it must reside, distributes execution to where learning occurs fastest, and builds feedback into the system so that improvement is continuous rather than episodic.

This book has argued that speed is not simply a matter of efficiency. It is a moral, operational, and strategic imperative. Speed compresses adversary decision cycles, reduces the window in which capability gaps can be exploited, and strengthens deterrence by signaling adaptability rather than static strength. It also serves a deeper purpose: an adaptive capability enterprise reduces the likelihood of conflict by sustaining credible advantage, and when conflict does occur, it reduces the burden placed on those who must fight.

C2G is not an argument for reckless acceleration. It does not abandon rigor, safety, or discipline. Instead, it recognizes that risk can only be managed, not eliminated. It also recognizes that unmanaged delay is itself a form of risk. By separating Effects from solutions, enabling multiple solution pathways, and integrating testing, learning, and tradeoff and termination authority into the process, C2G allows the force to move quickly without wagering everything on a single, irreversible decision.

C2G actually builds upon what already exists. The military must use caution not to "break itself to fix itself." The institutions, authorities, and organizations that sustain the force today are the foundation of C2G. Leaders, staff, Soldiers, industry, laboratories, academia, and the organic industrial base all play essential roles. C2G aligns those roles around shared Effects, clarifies ownership and responsibility, and provides the connective tissue required for the enterprise to function as a system rather than a collection of stovepipes.

History offers a consistent lesson. Periods of stability are sustained not by static dominance, but by the ability to adapt faster than competitors. The Pax Americana was underwritten by overwhelming capability, but that capability was generated in an environment that allowed time. That margin no longer exists. The modern contest, while absolutely leveraging mass where able, is still more about adaptation speed. It is an arms race of learning, integration, and execution.

The lesson from nature is instructive without being metaphorical. Systems that fail to adapt are eventually overtaken—not because they were weak, but because they were slow to adapt. Military capability is subject to the same pressure. Dominance without adaptability is temporary. Power without renewal is fragile. C2G institutionalizes adaptation so that learning becomes deliberate rather than accidental.

This book acknowledges that C2G cannot eliminate uncertainty or guarantee success. However, it offers something more durable: a framework designed to endure change and manage risk. C2G is built to absorb new technology, new threats, and new lessons in an ever-changing world without collapsing under its own complexity. In that sense, it is not an end state, but a way of operating.

The choice facing the defense enterprise is how to handle change. Will it shape the future deliberately or will it be a byproduct of external events?

C2G offers a way to manage adaptation and to generate capability at the pace demanded by the world as it is, not the world as it once was.

In an era where advantage belongs to those who adapt fastest, C2G is not merely how we build capability—it is how we choose to remain relevant.

GLOSSARY OF KEY TERMS

A

Acquisition Transformation

A Department of Defense initiative that shifts the acquisition enterprise toward portfolio-based management, rapid adaptation, and modular capability delivery. It emphasizes speed, integration, and iterative development rather than linear programs of record.

Adaptive Advantage

The ability to adjust faster than adversaries through learning, iteration, integration, and rapid decision-making. A core principle of C2G.

Auftragstaktik (Mission-Type Orders)

A historical German military philosophy emphasizing decentralized execution guided by clear intent. Provides doctrinal roots for modern mission command and informs the decentralized "limbs" approach in C2G.

C

C2G (Combat Capability Generation)

A framework that enables a unified, Effects-driven approach to transforming operational need into rapidly delivered, continuously improving military capability.

Capability

A means of achieving a desired military Effect. In C2G, capabilities are subordinate to **Effects**, not the other way around.

Capability and Effects Registry (CER)

The central, living "brain" of C2G. A repository of required Effects — not systems — that ensures unity of mission, prevents duplication of effort, and guides resource alignment.

Capability Portfolio Executive (CPE)

A leader responsible for synchronizing resources, programs, and decisions within a portfolio of capabilities. Works with PAEs to align solutions to CER-defined Effects.

CER Alignment

The process of ensuring all concepts, proposals, contracts, and development efforts trace directly to an Effect in the CER.

CITC (Combat Innovation & Transformation Cell)

A cross-disciplinary organization that translates Effect requirements into innovation pathways. A collaborator — not a gatekeeper.

Commercial-Off-the-Shelf (COTS)

Commercially available items that can be adopted with minimal modification to fulfill military needs.

Commercial Dual-Use Technology

Technology that has simultaneous civilian and military applications. Includes both commercial innovations adopted for defense and military technologies later adapted for civilian use.

Commercial Industrial Base

Privately owned, market-driven companies and supply networks that develop, produce, and sustain goods and services primarily for civilian markets, but which may be leveraged for military use through adaptation, contracting, or surge mobilization.

Competitive Overmatch

The sustained ability to out-adapt, out-integrate, and out-think adversaries. Emphasizes quality, speed, and intelligence, not mass alone.

Configuration Control

The discipline of maintaining accuracy, traceability, and consistency across technical documentation, designs, and program decisions.

Core Truth

A summarizing principle placed at the conclusion of each chapter to reinforce the essential message.

D

Data Rights & Intellectual Property (IP)

Legal and contractual privileges governing access to technical data. C2G promotes balanced protection for innovators and necessary access for the government.

Decentralized Execution

The concept that operational actions and decisions are executed at the lowest capable level while aligned to centralized intent.

Distributed Cognition

A system in which multiple nodes (units, labs, industry) perform adaptive decision-making while remaining aligned to common purpose — analogous to octopus or starfish biological systems.

Doctrine

Formalized principles that guide military action. In C2G, doctrine becomes adaptive and Effect-centric rather than static or prescriptive.

DOIB (Department of Defense Organic Industrial Base)

Government-owned or government-operated industrial facilities. In C2G, the DOIB is positioned as an alternate innovation and supply chain engine — not a secondary one.

DOTmLPF-P

Doctrine, Organization, Training, materiel, Leadership, Personnel, Facilities, Policy. In C2G, it becomes an operating system for transformation rather than a post-decision checklist.

Dual-Use Flow

The bi-directional movement of technology between civilian and military sectors.

E

Effect

A desired change in the environment or operational situation. Effects are the unit of value in C2G.

Effect Ownership

In C2G, Effects are owned by missions, not by branches, tribes, or functional communities.

Effects-Based Model

A planning and decision framework that organizes capability, resources, and actions around clearly defined operational Effects rather than predefined platforms, solutions, or inputs, enabling adaptation as conditions, threats, and opportunities change.

Embedded Personnel

Individuals — often Soldiers or operators — placed inside innovation, design, or development environments to provide ground truth and human context.

F

Facilities (Within DOTmLPF-P)

Physical infrastructure required to support new capabilities. In C2G, facilities are modular, adaptable, secure, and cyber-resilient.

Feedback Loop

The iterative cycle where data from execution or informs continuous refinement in the CER and C2G governance.

G

GOTS (Government-Off-the-Shelf)

Government-developed products made available for broad use without commercial development cycles.

Governance Loop

The C2G process that ensures alignment across CER, CITC, PAEs, CPEs, and execution "limbs."

H

Hobart's Funnies

Specialized armored vehicles developed in WWII demonstrating Soldier-driven innovation under urgent operational conditions.

Human-Centered Capability

A principle that humans — not systems — are the core reason capability exists. C2G prioritizes operators, ethics, and humanity above technological novelty.

I

Innovation Surge

The rapid scaling or modification of a capability based on need. Enabled by flexible contracting and modular design.

Iterative Development

Continuous refinement of systems or concepts based on data, testing, and operator feedback.

Industrial Base (National)

The total combined capability of commercial companies, defense contractors, universities, and labs that support national defense.

Instruments of National Power

The integrated diplomatic, informational, military, economic, intelligence, legal, and industrial capabilities through which a nation produces strategic Effects, deters conflict, and preserves stability.

J

JCA (Joint Capability Areas)

A high-level framework used to categorize capability domains. Complementary to CER but not a substitute for Effect-based alignment.

Junior Soldier Insight

Feedback derived from those closest to frontline friction, often revealing constraints unseen by senior leaders or designers.

L

Law of Armed Conflict

The legal constraints embedded in military operations that ensure force is used lawfully, proportionally, and discriminately, preserving legitimacy and accountability while enabling mission accomplishment.

Learn Fast

A core C2G discipline: rapidly extracting actionable insights without glamorizing failure.

Limited User Assessment (LUA)

A structured operational exposure allowing prototypes to be evaluated by representative users.

Limited User Test (LUT)

A formal test event with controlled risk acceptance to assess operational effectiveness.

Limbs of Execution

Distributed nodes of action (units, labs, industry, DOIB) that execute capability generation while aligned to centralized intent.

M

Manhattan Project (Reference)

Parallel development illustrating how massive capability leaps occur through concurrent exploration.

MDMP (Military Decision-Making Process)

A structured process for analyzing missions and decisions. C2G uses MDMP as a cognitive discipline.

Middle Tier Acquisition

An acquisition pathway used to quickly generate operational Effects through rapid prototyping or fielding, prioritizing speed, learning, and adaptability over traditional milestone sequencing.

Mission Command

The philosophy of leadership enabling disciplined initiative within commander's intent.

Mission-owned

Effects are owned by the operational commander accountable for achieving the outcome in real missions—typically a Combatant Command, Joint Task Force, or Service Component commander.

Modular Contracting

Contracts structured to support incremental development or rapid adaptation.

MOSA (Modular Open Systems Architecture)

A design and acquisition approach that requires systems to be built from modular, interchangeable components using open, well-defined interfaces to accelerate modernization.

N

Non-Developmental Item (NDI)

Equipment available that requires minimal adaptation to meet military needs.

National Power (Elements of)

Information, military, economic, diplomatic, and infrastructure resources that underpin national strategy. C2G influences several simultaneously.

O

OODA Loop (Observe, Orient, Decide, Act)

A continuous decision cycle that explains how individuals and organizations adapt to change and gain advantage through faster, more effective decisions.

Organic Industrial Base (OIB)

Comprised of the U.S. Commercial Industrial Base (CIB) and the Defense Organic Industrial Base (DOIB).

Other Transaction Authority (OTA)

A contracting mechanism designed for rapid prototyping and innovation without traditional FAR constraints.

Overmatch

The ability to maintain superior capability relative to adversaries across multiple domains simultaneously.

P

Parallel Development

Running multiple technological approaches simultaneously to reduce strategic risk.

Pax Americana

The condition of global stability enabled by sustained U.S. leadership and deterrence, in which credible military capability, economic influence, and institutional norms reduce the likelihood of large-scale conflict.

POI (Point of Impact) Risk

How easily the enemy can break your plan by hitting one thing.

Portfolio Acquisition Executive (PAE)

Senior leader responsible for integrating acquisition strategies across Effect portfolios.

Prototype Pathway

A structured approach to testing, evaluating, and scaling new capability.

R

Redundancy (Strategic)

Multiple viable pathways to achieve an Effect. Encouraged.

Redundancy (Duplication)

Unintended overlap of effort. Discouraged.

Risk Acceptance

Intentional acknowledgment of operational and testing risks, with formal boundaries.

Risk-Managed Decisiveness

A leadership trait emphasizing timely, informed decisions under uncertainty.

Risk Management

The deliberate evaluation and handling of risk as a variable to be understood, bounded, and employed in pursuit of desired Effects.

S

Scalability (Within C2G)

The ability to quickly expand capability from pilot to enterprise level.

Security Vetting

Procedures that ensure commercial or foreign-developed technologies do not introduce cyber or supply-chain vulnerabilities.

Starfish Analogy

Represents distributed resilience and local autonomy in execution limbs.

Surge-Ready Contracts

Contract structures that allow rapid increases in production or adaptation.

T

Testing (Operational)

Evaluation methods including LUA and LUT designed to validate capability under realistic conditions.

Transformation in Contact

The disciplined adaptation of capability, tactics, or systems while forces are operationally engaged, enabled by feedback, authority, and governance rather than the suspension of oversight or accountability.

Two-Way Dual-Use Flow

See dual-use: describes the reciprocal movement of innovation between military and civilian sectors.

U

Universal Capability Generation (Future Concept)

The application of C2G principles to other sectors such as healthcare, infrastructure, or disaster response. Intentionally saved for a future volume.

W

Who Dares Wins

The guiding ethos of C2G: disciplined courage, ethical decisiveness, and adaptive leadership. A phrase that originated with Sir David Stirling, the founder of the British Special Air Service in 1941.

V

Valley of Death

The systemic failure point where capability with demonstrated potential is lost due to delays, misaligned incentives, or lack of transition authority—rather than technical inadequacy.

REFERENCES

Department of Defense. *DoD Directive 5000.01: The Defense Acquisition System.* September 9, 2020.

Department of Defense. *DoD Instruction 5000.02: Operation of the Adaptive Acquisition Framework.* January 23, 2020 (as amended June 8, 2022).

Department of Defense. *DoD Instruction 5000.80: Operation of the Middle Tier of Acquisition.* December 30, 2019 (as amended).

Department of Defense. *DoD Instruction 5000.85: Major Capability Acquisition.* August 6, 2020.

Department of Defense. *DoD Instruction 5000.87: Operation of the Software Acquisition Pathway.* October 2, 2020.

Department of Defense. *DoD Instruction 5000.88: Engineering of Defense Systems.* November 18, 2020.

Department of Defense. *DoD Instruction 5000.89: Test and Evaluation.* November 19, 2020.

Department of Defense. *DoD Instruction 5000.97: Digital Engineering.* December 21, 2023.

Department of Defense. *DoD Directive 7045.14: The Planning, Programming, Budgeting, and Execution (PPBE) Process.* January 25, 2013.

Chairman of the Joint Chiefs of Staff. *CJCSI 5123.01I: Charter of the Joint Requirements Oversight Council and Implementation of the Joint Capabilities Integration and Development System.* January 11, 2021.

Chairman of the Joint Chiefs of Staff. *CJCSI 5123.01I, Enclosure E: Joint Capability Area Management Plan.* 2021.

Department of Defense, Office of the Under Secretary of Defense for Research and Engineering. *Modular Open Systems Approach (MOSA) Program Guidance.* 2019.

Government Accountability Office. *Weapon Systems Annual Assessment.* 2023.

National Institute of Standards and Technology. *NIST Special Publication 800-53: Security and Privacy Controls for Information Systems and Organizations.* 2020.

National Institute of Standards and Technology. *Secure Software Development Framework (NIST SP 800-218).* 2022.

Boyd, John R. *A Discourse on Winning and Losing.* 1987.

Clausewitz, Carl von. *On War.* Princeton University Press, 1989.

Sun Tzu. *The Art of War.* Oxford University Press, 1963.

McChrystal, Stanley. *Team of Teams: New Rules of Engagement for a Complex World.* 2015.

Taleb, Nassim Nicholas. *Antifragile: Things That Gain from Disorder.* 2012.

AUTHOR BACKGROUND

Brian C. Stehle is a retired U.S. Army Colonel, Apache helicopter pilot, and acquisition professional with more than 27 years of service. His career spans operational aviation, major defense program management, enterprise transformation, and strategic modernization planning across the Army and Joint Force.

Stehle's work focuses on redesigning how the nation generates combat capability—from industrial mobilization and acquisition reform to organizational design and decentralized execution. His perspective bridges operational military experience with the realities of modern defense acquisition and national-scale capability development.

He is also the author of *One Day Off*.

Made in the USA
Coppell, TX
05 November 2026

11315634R00104